What people are saying about

The Circle of the Snake

Whether it is dissecting the u_____ ____ ___ ____ tech gurus, revealing the political econo_____ ___ __ nostalgia industry, or uncovering exploitation in the global division of digital labor, *The Circle of the Snake* is just the guide we need to navigate, resist, and transform the online world. In precise, accessible prose, Grafton Tanner offers a much-needed vision of democratic alternatives to surveillance capitalism.
Vincent Mosco, author of *The Smart City in a Digital World*

In an age of platform capitalism in which state power is transformed by Big Tech conglomerates who exercise increasing power over our movements, thoughts and ideologies, Grafton Tanner finds a way out via critique. Our particular brand of capitalism is best understood through a nostalgia industry which it uses to sustain itself, Tanner shows. Everything from popular culture, politics and music to fashion, TV and videogames can be seen as bound up with forms of nostalgia inherently connected to our economic conditions. By understanding nostalgia, Tanner lays bare our ideologies and confronts not only the historical development of capitalism but the particular condition of today's digital corporate world.
Alfie Bown, author of *The PlayStation Dreamworld*

The Circle of the Snake

Nostalgia and Utopia in the Age of Big Tech

The Circle of
the Snake

Nostalgia and Utopia in the
Age of Big Tech

Grafton Tanner

Winchester, UK
Washington, USA

JOHN HUNT PUBLISHING

First published by Zero Books, 2020
Zero Books is an imprint of John Hunt Publishing Ltd., No. 3 East St., Alresford,
Hampshire SO24 9EE, UK
office@jhpbooks.com
www.johnhuntpublishing.com
www.zero-books.net

For distributor details and how to order please visit the 'Ordering' section on our website.

ISBN: 978 1 78904 022 7
978 1 78904 023 4 (ebook)
Library of Congress Control Number: 2019948943

A CIP catalogue record for this book is available from the British Library.

Design: Stuart Davies

UK: Printed and bound by CPI Group (UK) Ltd, Croydon, CR0 4YY
US: Printed and bound by Thomson-Shore, 7300 West Joy Road, Dexter, MI 48130

We operate a distinctive and ethical publishing philosophy in
all areas of our business, from our global network of authors to
production and worldwide distribution.

Contents

Also by Grafton Tanner

Babbling Corpse: Vaporwave and the Commodification of Ghosts
(ISBN 978-1-78279-759-3)

To Anna,
The world is brighter with you in it

Se acabo el tiempo de la esclavitud.
- *Versatronex employee Joselito Muñoz*[1]

Surfing has already replaced the older sports.
- *Gilles Deleuze*[2]

Who dares dissent from the gospel according to Silicon Valley?
- *Mark Fisher*[3]

Introduction

The Myth of Digital Utopia

Seventeen years before employees at the Foxconn City industrial park in Shenzhen, China committed suicide by throwing themselves from factory buildings, workers at Versatronex Corporation in Silicon Valley organized a protest. Sick of working with toxic chemical baths without health insurance, women factory workers launched a hunger strike while men camped in tents outside the headquarters in solidarity. Most were Central American, Mexican, and Filipino immigrants. They were paid paltry wages and denied information about the dangers of the chemicals involved in manufacturing electronics, for clients like IBM and Digital Microwave Corp. The strike ended in November 1992, and by January 1993, Versatronex was closed. Instead of meeting the demands of workers by opening safer factories in the U.S. and paying respectable wages, Silicon Valley outsourced its manufacturing overseas.[1]

The digital technologies of the twenty-first century can only exist thanks to this kind of outsourced labor. The relative invisibility of the tech supply chain is part of the ruse; American consumers do not see where smartphones come from. They do not see the conflict zones where coltan is mined to be used in electronic devices, or the sweatshops in which digital products are manufactured. The latest technologies arrive instead in pristine condition, as if delivered from on high. New tech developments are displayed on vast stages by their designers, who act as sorcerers demonstrating the mighty transformative power of their products. Currently our technologies are made by people, but with advancements in machine learning and automation, that will likely change. If Foxconn's specialized robots – 'Foxbots' – outpace human labor, they will make our

1

smartphones instead.[2]

In the meantime, workers in China, South Korea, Indonesia, the Philippines, Vietnam, Thailand, and other countries around the world are the ones producing consumer electronics, oftentimes working in disturbing conditions to meet outrageous demands. In 2007, when Steve Jobs decided at the last minute to make the iPhone's screen out of glass, 8,000 workers at a factory-built dormitory in China were awakened in the middle of the night. They were each given tea and a biscuit to labor around-the-clock, assuring Jobs' deadline would be met.[3] When Foxconn employee Sun Danyong was accused of stealing the prototype of a new iPhone in 2009, he was beaten and his apartment searched by Foxconn's security team. In July of that year, he jumped from the twelfth floor of his apartment building.[4]

Similarly, we don't see the human labor that ensures the largest social media companies in the world run smoothly. Content moderators scour the filth and refuse uploaded to Facebook, Instagram, and Twitter and do so earning undignified wages in workplaces where productivity is constantly tracked. The sheer volume of questionable posts on these sites combined with unreasonable expectations has engendered a public health crisis gone largely unnoticed. Moderators suffer from mental illnesses, and over time employees often find themselves numb to the appalling things they've seen. Some start believing in the very conspiracies they are tasked with removing.[5] Part of the job of a content moderator involves chasing down and deleting the extreme emotional reactions users post online in order to give the appearance that social media are open platforms for measured public debate. But the algorithms that structure social media often facilitate a hostile kind of debate punctuated with quick flashes of emotions.

In 2013, the same year that psychologist Ethan Kross published a paper revealing that Facebook undermines wellbeing, scientists at Beihang University proved that the emotion spread most widely through social networks is anger. Joy, sadness, and disgust, which the scientists also mapped, could not compete.[6] Even in 2013, this finding came as no surprise. Anyone who has spent considerable time on social media might attest that it can be a hostile place. After all, anger garners clicks, which is a boon for corporations whose primary business goal is capturing your attention.

Because it's a scarce commodity, there's only so much attention to go around, which means advertisers and corporations must fight to secure it. Once they do, they must maintain it to increase profits. In an attention economy, some good or service is provided to a person in exchange for their attention. Securing one's attention is achieved by giving someone something they want: a news story, the definition of a word, a weather forecast, or an update on what a friend is doing. While people read through this information, ads are shown to them. Some ads appear native to the site; they don't resemble the traditional advertising we may be used to seeing. When consumers pick up on what ad companies are doing, then the burden is on those companies to figure out more subtle ways to influence.

If people are shown the most relevant, up-to-date information about their friends, family, co-workers, and the wider world, they are more likely to keep reading and refreshing the page. At the same time, people tend to respond to information tailored to their interests. If companies know the products you like and show them to you online, you'll spend more time scrolling. The goal is to harvest as much information from a given person as possible in order to recommend and even predict that person's desires.

Much of the machinations of the attention economy are well outlined, and many people online understand they are being advertised to. How your information is used and what happens psychologically in an attention economy is now beginning to show. Some criticize the attention economy for not only addicting users on digital products but also influencing politics and corroding democratic values. Yet the attention economy didn't suddenly spring into existence. It is a result of decades of neoliberal logic, free-market fanfare, and widespread deregulation.

Disruption and Deregulation

Unregulated market forces tend towards monopoly. It is no surprise, then, that Google and Facebook own almost all Internet revenue, and they are often applauded for their financial success. They're also praised for disrupting the older ways of capitalism, and since we live in a neoliberal culture of competition, this kind of market disruption is considered natural. The fittest, as it's been said, survive, so if a company figures out a way to subvert the system, they are commended for their business acumen.[7]

The corporate world glorifies Big Tech for bucking the system and overthrowing the business status quo, which perpetuates a popular cultural narrative that shows up in movies like *Pirates of Silicon Valley* (1999) and *The Social Network* (2010). This narrative stars a particular kind of character: the young male clad in casual dress who creates a tech company in his basement or college dorm room. He is the twenty-first century Ragged Dick, and from his hard work, he climbs the social ladder, endures hardships, turns failures into triumphs, and eventually achieves considerable wealth. The older guard regards him as honest, maybe a little nerdy; at times, he acts pathological. But as long as he works hard and makes money, who cares if he attends board meetings in a hoodie? He's seen as an outsider, or even a prophet, who saw the coming computerized world and adapted more quickly than the rest of us. Now that he paved the way, perhaps we too can start our own successful tech companies and be like him.

But these tech disrupters, like Jeff Bezos, Mark Zuckerberg, Larry Page, and Sergey Brin, aren't freaks of capitalism. They haven't subverted any system. Rather, they and other Valley technocrats merely acted unilaterally, free from the confines of pesky regulation. Free-trade and consumerist ideologies gave Big Tech the resources to manufacture and the markets to sell digital technologies. After privatizing nearly everything, neoliberal ideologues aimed their weapons at our attention. It was no longer enough to pleasure the American people with cars, refrigerators, and clothing. Once it exported the jobs of the working class to other countries and hammered the final nail in the regulatory coffin, neoliberalism set the stage for Big Tech's entrance.

Much of the labor it takes to power Big Tech is largely unseen by Western eyes. Rather we are shown fit, smiling, young men on stages, delivering speeches accompanied by visually appealing slideshows. They are our tech leaders, the geniuses like Steve Jobs and Bill Gates who conjured into being these fantastic inventions that quickly connected everyone across the world at the dawn of the twenty-first century. When they arrived, they were like gifts, and their developers were heralded as demigods. Older digital migrants made every effort to remind younger digital natives how hard life was before these sublime technological gifts. The natives have known no other world.

I have lived as something in between – not quite a migrant but more native than others. The first years of my life were spent largely without a home computer. In the early 2000s, this was already difficult, as much of my schoolwork had to be typed in word processors. Most weeknights, I would work late in my mother's classroom, which was stocked with computers. By the time I completed high school, my family owned a secondhand computer, and for the first time ever I experienced what it was like to socialize online. Some friends encouraged me to create a profile on this website that had until then been offered only to college students. It was called Facebook.

At that point, one kid in my school owned a smartphone. The

rest of us had convinced our parents to buy us flip phones and other early mobile devices. Because the student who owned a smartphone came from a wealthy family, we associated such devices with affluence and ostentation. Surely, no one needed a phone that could do everything. I regarded them as fads that either gearheads or the rich would purchase, and I presumed they would never enter the mainstream. By the time I matriculated into college in 2009, I knew I had been wrong.

Digital Mythology

It's growing increasingly difficult to remember a time before Big Tech's ascension. Some have attempted to do so and failed considerably, portraying the late twentieth century especially as a simpler, more spontaneous time. Some like to think the people of history dreamt of Big Tech and yearned for its magical devices to solve the daily problems of life in the pre-digital era. The former is the view of the nostalgic sufferer. The latter, of the digital utopian. Both are the dominant subject positions of our time.

One of the most dangerous and powerful myths of the present century is that digital technology can solve any problem and, thus, bring about utopia. With open markets and open communication lines, people could freely exchange ideas without the middlemen of previous media. Peer-to-peer sharing would lead to a democratization of art and information. Early Internet pioneers considered the future to be one of endless possibilities. The run-up to the millennium was defined by futurism, cyberpunk, and accelerationism. Reacting against primitive leftism, which envisioned a woodsy utopia complete with campfire songs and organic foods, the technophilia of that time led many to believe the Internet would knock down the last known pillars of oppression. And the technology-to-come, such as virtual reality, would allow humans limitless forms of expression.

Web 2.0 has yet to create any kind of functional utopia. It never will. Instead, it serves as a machine for the circulation of anger, as the researchers at Beihang University found. But there is another emotion that disseminates through the channels of the Internet, one far more complex and difficult to map: *nostalgia*. In the digital age, nostalgic representations of the pre-Internet era pop up everywhere, from streaming series to movies, music to fashion. It seems the more we become tethered to mobile devices and imbricated within social networks, the more we yearn for a time before Big Tech.

It isn't that we want to be completely rid of digital technologies, but rather that we now have unfettered Internet access to much of the popular content from the decades before Big Tech, content that appears to portray the past as simpler, less busy, less anxious, and more prone to chance. Without smartphones, the citizens of the past represented on screens seemed to live fully in the moment without the distractions of Twitter or Instagram. One had to know how to operate a camera, remember phone numbers, read a paper map, stay up for the evening news or a television show when it aired, wear an analog wristwatch, wait for the arrival of paper mail, catch a favorite song on the radio, set an alarm clock, purchase goods from a brick and mortar, and live with the painful pangs of boredom and unknowing. As media converged, these ways of life under capitalism diminished, but thanks to a trove of cultural artifacts held over from previous time periods, we can catch glimpses of how people used to behave before digital proliferation. And because the Internet collects these artifacts for anyone to see, we are only a click away from feeling instant nostalgia.

However, representations of the past, especially those from popular culture, do not tell the whole story. When they first aired, whether in the 1950s or the 1990s, these corporate-approved tales left out the nasty bits of history. Although sometimes political messages are smuggled into popular representations, pop culture on the whole is a vehicle for advertising. For this reason, it is a site of struggle. Content creators, consumers, and prosumers negotiate the meanings of popular texts, which react to society even as they create social realities. Yet much of our nostalgia today – what I call pre-Recession nostalgia – appropriates representations from the mainstream pop culture of the past, advancing an agenda that served corporate and sometimes imperialist interests.

What we are nostalgic for today is the history told by media corporations. When we think life in the past was as simplistic as *Leave It To Beaver* or *The Goonies*, we risk pining for a suffocating social reality that's whitewashed, normative, and patriarchal, as many of the mainstream narratives of the late twentieth century were. But those most desperate for a return to the misremembered golden days of yesteryear might do whatever it takes to get us back there.

This turn towards nostalgia in our time is not surprising; nostalgia peaks during periods of social, political, or even personal unrest. Reeling from the 9/11 attacks and the global economic meltdown in 2008, many in the West were already disillusioned when businessman and TV personality, Donald J. Trump, announced his presidential candidacy in 2015. His rhetorical tactic of 'truthful hyperbole' resonated widely with disenchanted poor whites, as well as fringe groups: nativists, protectionists, and white nationalists. He appealed to citizens whose way of life had declined over recent years, who did not recover as soundly from the Great Recession, and who saw a deep-seated resentment towards white, working-class Americans in establishment politics.

But the nostalgia of today isn't simply a reaction to a technologized, globalized world. It also flows directly from the algorithms that power late capitalism.

Controlling the Past

Big Tech has helped foster not only an attention economy but also a nostalgia industry. A branch of the culture industry, the nostalgia industry sells simplistic ideas about history that often have damaging effects on society. From *Stranger Things* to 'Make America Great Again,' nostalgia has circulated more widely in Western culture since the Great Recession, due in part to the algorithms that structure Big Tech's products. At a time when historical literacy is crucial, when old prejudices are starting again to percolate into the present, Big Tech's algorithms resist any attempt to exit the feedback loop of amnesia. There have been attempts to explain *why* we are seemingly more nostalgic now than ever before, but insufficient attention has been paid to Big Tech's complicity in our collective longing for the past. For those readers who are curious as to why everything old is new again, this book will provide in-depth backstories and critical interpretations.

It seems paradoxical: how could some of the most complex technologies ever invented induce nostalgia for the past? There are a few possibilities. The first is that Westerners are expected to always be 'on' now more than ever before. The rationality of neoliberalism, which promotes competition and entrepreneurialism, has left many stressed, depressed, and anxious. Digital devices and social media exacerbate these feelings, which in turn are further aggravated thanks to a precarious economy. For many, living and working have become quite difficult – for some, even impossible. Societies experiencing instability have a natural tendency to retreat into the past, when things were supposedly simpler and easier.

Second, the Internet itself is a portal into the past. As I argued in *Babbling Corpse*, the Internet gives individuals experiencing extreme stress access to nearly all of human history.[8] Armed with this tool, many find it preferable to escape into versions of the past offered online.

And finally, the structures of social media and online advertising encourage nostalgia to circulate. Recommender systems and predictive analytics – the very tools that allow our contemporary media to function – zero in on quick reactions, such as a flash of anger or a swell of nostalgia. These reactions are noted by algorithms, which then make recommendations based on them. Artists and producers, struggling in a culture industry decimated by piracy, YouTube, and streaming services, create media they know will be picked up by the algorithms. The result is a nostalgic feedback loop wherein old ideas travel round.

If nostalgia were simply an emotion of remembrance, then perhaps we wouldn't be witness to the rise of nativist movements, bent on reclaiming a version of the past that might not align with the goals of democracy. Theorized for centuries, nostalgia translates as 'the ache to come home.' For some, 'home' is remembered fondly as a safe place that instills good feelings. For others, remembering 'home' is so painful that it must be returned to by any means necessary. When nostalgia metastasizes, afflicted individuals may take extreme measures to alleviate the ache. For them, nostalgia is a burning desire to control the present state of things by reclaiming what is thought to be lost: old ways of life and, along with them, longstanding prejudices. Considered in this way, nostalgia can be an emotion of control and cannot be properly written about without understanding how social control works in contemporary Western societies.

In a control society such as ours, the state and multinational corporations monitor individuals with great precision. Advertisers, in turn, market to us on the most granular levels in an attempt to predict our future desires. Increasingly complex digital technologies give users the kind of control once reserved only for rulers and elites in older societies. In short, the desire to revive the past dovetails with the dream of total control. In the process, privacy is eradicated.

Very often, digital technologies are positioned as legitimate means of gaining control over things. The prevailing myth is that Silicon Valley's inventions will fix fundamental human problems by delivering the power of control into the hands of users. This myth appears time and again throughout the twenty-first century, from the mouths of the technocrats themselves to films, streaming series, and memes. And one of the promises of digital technology is that it will free us from the shackles of history and relieve the intense cultural nostalgia for previous time periods.

This book is about Big Tech's promise of digital utopia, a world in which every human dilemma is solved by digital technology and everything is under control. In a digital utopia, all of life is managed from above by tech overlords. Everything is perfect; you can live in whatever time period you want. You can be whatever you want to be. This world doesn't exist, but Big Tech promises us that, if we give it absolute power, it will bring it into existence. And there are several popular Western narratives that circulate this myth. Few of them shed light on its manifold dangers – namely, the obliteration of privacy, the explosion of mental health disparities, the dream of social control, and the siren song of nostalgia.

One cannot remember without also forgetting. Those afflicted with pre-Recession nostalgia also present with symptoms of amnesia. Filtering memories out of one's mind is a natural part of remembering, and this can be a dangerous move for a society to make, especially when we forget the atrocities of our history. But even if we do remember, even when we know all too well the world as it once was and make every effort to thwart societal regression, we have to reckon with the interests of elites, who may have much to gain from the circulation of pre-Recession nostalgia. Imagining a radical nostalgia – one crafted from memories of collective resistance, community organization, civil rights, and local politics – requires countering the neoliberal assault on the rights of people. It also means coming to terms with the myth of digital utopia told and re-told by Big Tech.

Chapter One

To Manage the Universe: On the Sublime Power of Digital Technology

I miss the cameras. They used to be heavier than us. Then they became smaller than our heads. Now you can't see them at all.
- *from* Holy Motors *(2012)*[1]

There's no unlit corner of the room I'm in.
- *Torres, 'Skim'*[2]

There was a time when the power to analyze others was not given to ordinary people. The ability to scrutinize the smallest details and interpret them was either reserved for trained individuals or impossible without an apparatus to aid in close inspection.

At the turn of the fifth century, some ten years after he heard the voice of God in his garden, Augustine of Hippo wrote *On Christian Doctrine* and claimed that the Bible, like the great classics of Greece and Rome, could be interpreted by experts trained in the art of exegesis. Centuries later in Vienna, Sigmund Freud put forth the theory that skilled therapists could interpret human dreams in much the same way, thus inventing the mode of psychological study called psychoanalysis. For both Augustine's Scriptural hermeneutics and Freud's psychoanalysis, a proficient critical mind was the tool needed to uncover the latent meanings in things. With the right training, one could draw out the truth and make judgments. But with the invention of photography in the nineteenth century, the power of interpretation was no longer limited to experts.

On the eve of the Second World War, cultural critic Walter Benjamin wrote that, thanks to the invention of photographic film, things around us were getting closer. Passing events suddenly became easier to isolate and inspect. The smallest gestures could be targeted and studied. And it was film, Benjamin believed, that facilitated the changing way people looked at things and understood them. Fittingly, he connected this shift to Freud's theory of psychoanalysis. 'Fifty years ago,' he wrote, 'a slip of the tongue passed more or less unnoticed.'[3] Photographic film simply took the power to probe behaviors a step further. One no longer needed to be a psychoanalyst to scrutinize the subtle elements that make us tick. Daguerre's portraits, Muybridge's galloping horses, and the Lumière brothers' train arrival were preserved for the eye to catch and the mind to study. In the dark of the cinema, we all become analysts.

Film was, and continues to be, liberatory. It blows apart space and time and shatters the barriers between people. Benjamin was struck by this power, and he knew that even in its infancy, film would totally alter human perception:

With the close-up, space expands; with slow motion, movement is extended. The enlargement of a snapshot does not simply render more precise what in any case was visible, though unclear: it reveals entirely new structural formations of the subject...The camera introduces us to unconscious optics as does psychoanalysis to unconscious impulses.[4]

Other technologies have brought things even closer since Benjamin penned those words. With each new invention, we take yet another step towards both the world around us and the inner worlds of each other.

By the middle of the twentieth century, television delivered this analytical power into the living room. Cocooned in the safety of home, drawn by the analog glow, families gathered round to peer through a portal to the outside world. Decades later, the Internet initiated the process of connecting people to information and each other. As an old century came to a close and the millennium dawned, we learned of a shadow world mirroring our own, a reality online that existed parallel to, and sometimes even eclipsed, the one offline. More than ever, the outer world crept closer to the monadic lives we had been living. Many believed that the connective power of the Internet would usher in a utopian reality, where nothing would be left unknown and no person left alone.

In the first decades of the twenty-first century, we are connected like never before. We carry the tools of our analysis – smartphones, social media, wearable tech – with us everywhere we go. We analyze each other's speech and the behavioral slips of politicians. We watch closely the lives of celebrities, as the slightest facial tic can go viral. We dissect bodies and ideas with the clinical precision of surgeons. We take apart films, music, and art of any kind and share our critiques with others. We spy on people as we ourselves are spied upon. We track and monitor our every move. At last, everything is in sharp relief.

As things grow closer, the technology we use to analyze the world becomes normalized. Devices get smaller, more streamlined. Some technologies become institutionalized; suddenly, they appear at work and in the classroom. It seems that our machines will never be replaced, and then something else comes along and changes everything. When this paradigm shift occurs, entire industries disappear. Ideas and beliefs get re-routed or lost forever between the cracks. Some technologies persist or re-appear at some later juncture. They percolate through other points in time, weaker areas where the cultural fabric is worn thin. At moments when instability is high and nostalgia amplified, older technologies bubble up from the past and exist alongside digital contemporaries. The re-appearance of dated things may seem odd at first sight. But pretty soon they too become as normal as if they'd never disappeared.

As we analyze each other, so too are we watched.

Corporations monitor our purchases online and in-store. Our data is intercepted by third-party companies and sold to federal organizations like the NSA and FBI. The Five Eyes – made up of the United States, the United Kingdom, Australia, Canada, and New Zealand – comb the world with precision, leading the way in global surveillance measures.

The price of unfettered connection and total magnification is surveillance. While major public and private institutions watch us, we watch back. Instances of police brutality are filmed by passersby on smartphone cameras. The words and actions of powerful leaders are tracked astutely. Bigots and oppressors are held accountable. Cultural linchpins are ousted from positions of power. It is quite difficult to predict who might fall under the watchful eye of the people, who wield enormous influence thanks to user-generated communications platforms like Twitter and Reddit. Protest movements seemingly explode without warning. The cries for justice grow louder online and off.

And we don't just watch; we watch closely. Some scour social media feeds for damning clues to discover who among them is an oppressor. Obvious cases of vitriolic intolerance are noticed widely. The perpetrators of prejudice are roundly criticized and banished from public acceptance. In an unjust twist, some exit the public stage with massive severance pays, such as Fox News pillars Bill O'Reilly and Roger Ailes, both accused of sexual harassment and who both received millions upon leaving their positions.[5]

Even as irredeemable figures are ousted from the spotlight, in our global culture of surveillance, a major casualty is privacy. Our data trails snake through corporate and federal institutions while our peers watch us through the lens of social media. Like the rattled prisoners in Bentham's panopticon, a circular prison with a watchtower at its center, we are unsure whether or not we are being watched. So we behave as if we are always on the stage, as if someone is always listening in.

The feeling that life is a stage is nothing new. From Erasmus to Shakespeare, the idea that humans are mere actors performing roles for ourselves and each other has persisted, so much so that sociologist Erving Goffman coined a theory in the 1960s to describe it. Goffman claimed that humans play certain roles and don faces to present to others who they think they are. His theatrical approach to reality maintained that other people play their own parts as audience members or as other actors and that the dreadful fear of exposure – that perhaps someone may stumble backstage and see an actor without her face – keeps the play of life going. Everyone, according to Goffman, has a role, feels good when others validate their role, and ensures the show goes on without a hitch. Embarrassment and aggression arise when faces are unmasked and the play is disrupted.[6]

In the decades since Goffman created face theory, the backstage has bled onto the stage itself and into the audience. Those in the crowd want to know what's backstage. There is no longer any proscenium, no barrier between actor and audience. There is only a smooth, flat stage extending in all directions at once, and the fear that once gripped the actors is heightened tenfold. There is no backstage to hide.

Whatever privacy left is routinely violated. Sensitive information is leaked through encrypted channels, exposing injustices on both micro and macro scales. Monitory organizations expose the hypocrisy of liberal democracies by blowing the whistle on massive surveillance operations and human rights abuses. And in this age of 'communicative abundance,' as political scientist John Keane calls it, some individual public actors are regularly revealed to be charlatans.[7]

Behind Closed Doors

A decade after Goffman created face theory, as the Watergate scandal unraveled in the early 1970s and tape-recorded conversations became public news, President Richard Nixon believed he had done nothing wrong. Communication theorist Joshua Meyrowitz blames Nixon with a failure to grasp the enormity of the new technologies of his time.[8] Nixon portrayed himself as a leader in the old style – the kind of great figure that existed before television destroyed the last vestiges of a leader's mystery. He was unaware that what seemed private to him was ultimately public for the nation and the rest of the world. By failing to understand the problems with taping private conversations, Nixon had sealed his fate.

To support his interpretation of Nixon, Meyrowitz offers the example of a 'bull session,' or, as it has also been called, locker-room talk. Meyrowitz understands locker-room talk to be something boys do behind closed doors that, if brought into the open light, would be offensive and revolting.[9] After hearing then-presidential candidate Donald Trump caught on tape boasting about how he physically handles women, his public relations team and supporters wrote it off as mere locker-room talk, something boys just do in private and, therefore, nothing of real consequence. But his critics were appalled. How could a presidential candidate say this? Many believed that someone running for public office should not behave in such an offensive, puerile way. But had Harry Truman run for president in 2016, we would be equally appalled at his virulent intolerance and open support of the Ku Klux Klan.[10] It seems the U.S. has a long history of presidential candidates making shocking remarks in private.

Do we understand Trump's locker-room talk and Truman's private racism as Meyrowitz understands Nixon and his tapes? Is it permissible for men to engage in locker-room talk? Is it only permissible if no one hears? Or is everything public and open for analysis? Does it matter if privacy is breached if a bigot is thrown from a position of power? And when the NSA functions as a kind of Big Brother, why not hold it and other major players accountable? We are spied upon – why not spy back?

By spying back, we expose the mysteries of our leaders, for better and worse. Meyrowitz believes we long for another great leader, but such great leaders are made impossible by our technologies. 'There is no lack of potential leaders,' he writes, 'but rather an overabundance of information about them.'[11] He saw television as the destroyer of mystery, which is a key ingredient in making a great leader. By demystifying public figures, we reduce them 'to the level of the average person.'[12] This is particularly true for mainstream pop musicians. The mystery of Jeff Buckley, the showmanship of Freddie Mercury, the fluidity of Grace Jones and David Bowie, the glory of Prince – can there be another like them if we continue our demystification of public figures?

In the age of enhanced analysis and communicative abundance, we are coming to terms with our long-held separation of public and private, art and artist. In his 1981 book *The Political Unconscious*, Fredric Jameson wrote that no 'world-view...that takes politics seriously' could tolerate public figures without also accepting their private beliefs.[13] Considering the ideologies of public figures secondary or ignoring them altogether reinforces the notion that individuals (and art itself) can be decontextualized and that politics is not intimately bound with all things. This, Jameson notes, is the legacy of capitalism, which reduces ideas and people to things in themselves, shorn of context. History becomes trivia instead of what it really is – 'a nightmare.'[14]

We don't necessarily want our leaders to be average persons like us, even though we often enjoy hearing that famous celebrities eat the same fast food as regular people. But in the beginning of the twenty-first century, we carefully watch our public figures to ensure they do not commit an unconscionable act. Doing so helps to rid the public stage of bigots, even as it also threatens the last known walls of privacy. This is an inevitable tension that must be maintained as we open the door on the private lives of others.

There may not be, as Meyrowitz thinks, a plethora of potential leaders. The pathways toward public office are paved with the capital of crooks, weapons manufacturers, oilmen, Silicon Valley technocrats, and abusers. It is quite difficult for truly great leaders to take the stage when cronies direct the play. And it is a painful process to re-politicize what has been left apolitical for so long. Even as we attempt to reintroduce the political into social reality, to mend the split capitalism rent, surveillance becomes normalized and privacy, jeopardized. The analytical power of technology lets us hold others accountable, even as we act under the roving eyes of the surveillance state.

All Will Be Revealed

Accompanying the ability to analyze others more deeply is the increased desire to have things revealed. From daytime talk shows to reality TV confessions, American culture has been obsessed with extracting the hidden truth from within the minds of others, using interrogation techniques both informal (questioning) and enhanced (torture). The goal is to draw forth greater amounts of information, something our media allows us to do more casually. The combination of analytical media technologies with what Mark Andrejevic calls 'infoglut' has led to an uptick in conspiracy theories, meant to explain the growing complexity of the modern world.[15]

Unlike older technologies, the structure of the Internet facilitates conspiracy. From one hyperlink to the next, we fall through the endless abyss of the digital realm, never striking bottom. One thing leads to the next, and the next, and the next, until you end up somewhere far away from where you started. But a string of perhaps loosely connected – maybe even totally unconnected – ideas trails behind you. After much falling, the impression is that by design the pieces are all connected.

Initially, the Internet was praised as a freer way to encounter information. In the early 1990s, digital theorist George Landow saw hypertext as a liberatory reading strategy.[16] He embraced it as a common good and thought hyperlinks would emancipate readers from the prison of the fixed word, allowing them to flow freely between various information sources. But what Landow never imagined is the exhaustion that could come from the endless, rootless process of reading in this way. Not everyone wants to jump from point to point without any center. Infinity might not always be alluring.

Landow puts the agency in the reader when he writes, 'anyone who uses hypertext makes his or her own interests the de facto organizing principle (or center) for the investigation at the moment.'[17] But how much agency do we actually have when falling down the rabbit hole? What if we get lost in the hole where the center was? How much attention do we end up losing in a world where we must always multi-task and where reading itself is disrupted by one hyperlink after another? And who, by the way, is at the other end of those links? Are they professors like Landow providing critical articles meant to enrich your reading experience or a corporation redirecting you to its latest product you just have to buy?

Out of a hyperlinked society full of infoglut came scores of conspiracy theories, each one more hideous than the last. For years during his tenure, the neoconservative media blasted President Barack Obama, accusing him of being born outside the United States, readying death panels, and being the Antichrist.[18] A segment of the American population, fed on a diet of these and other equally ludicrous conspiracy theories, took to the polls to vote for Donald Trump, stopping the Democratic Party from destroying America any more than some thought it already had. It wasn't until Trump openly acknowledged certain conspiracies and backed them that all bets were off the table. '[O]nce mainstream political figures acknowledge conspiracy,' media theorist Jodi Dean warned in her 2002 book *Publicity's*

Secret, 'anything goes.'[19]

Many of the fringe conspiracy groups that gained popularity in the twenty-first century found refuge on Reddit, until the aggregator banned most of them.[20] Yet the ban only re-affirmed the belief among extremist groups that they have unlocked the truth. They have since congregated on various messageboards, chat channels, and on Voat, a Reddit alternative that provides an open platform for bigotry.[21]

But it is the popular website YouTube, which is visited by more than thirty million people every day, where conspiracy flourishes. Its recommendation algorithms poach the attention of visitors by showing them the most extreme videos uploaded by firebrands of all kinds. No matter how innocuous the topic, YouTube will likely direct you towards those subjects that are proven to hold your attention – conspiracies.

In an attempt to make sense of a complex world, conspiracists hold that all things can be explained. Surely with so much information, things must make sense. All that data must hang together to form coherent narratives. But the fact is, there's more information than anyone could collate. And the affairs of reality are too turbulent and chaotic to be mapped completely. Truth is often stranger than conspiracy. And even the most powerful people have little grasp on what is happening.

Technocratic Turncoats

The digital technologies of today facilitate a method of analysis unprecedented in the history of media. At the same time, they affect our lives on both micro and macro levels. They have been known to abrade wellbeing on an individual level and disrupt affairs at the social level.

Some of the loudest voices discussing the mental health effects of Big Tech have been ex-Valley technocrats.[22] In May of 2017, ex-Google employee and design ethicist James Williams presented a talk entitled 'Why (and How) to End the Attention

Economy,' during which he spoke at length about the impact Big Tech has had on society.[23] He claimed that billions today may be addicted to social media and that, at the time, he and other Valley software designers knew they were creating technologies tailored to hook people. With so many people addicted to social media, he argued, the results have been disastrous.

By the time Williams gave his talk, there had been murmurings of concern as some began questioning the mental health effects of social media.[24] But Williams went beyond these concerns and addressed the societal consequences of social media – mainly, in response to the events that had occurred just six months earlier.

For many, the unthinkable happened on Tuesday, November 8, 2016. After months of campaigning, name-calling, and divisive rhetoric, the 2016 presidential election resulted in a surprise victory for Republican candidate Donald Trump. Shortly after his win, reports leaked accusing Russian sources of buying advertisements through Facebook worth thousands of dollars.[25] It is suspected some ten million Facebook users and over half a million Twitter accounts were exposed to the Russian ads, which were allegedly designed to disrupt the democratic electoral process and spread 'fake news,' a process that many thought aided the election of Trump.[26] Suddenly, in the minds of many Americans, the divide between real and fake narrowed. The backlash against Facebook was tremendous enough to warrant a response from founder Mark Zuckerberg, who promised to 'make it a lot harder' for such interference to occur again, filter news articles out of users' feeds, and prioritize content like baby pictures and other 'meaningful interactions.'[27]

James Williams connected the dots during his talk. He tied the societal effects of social media to the neurological effects of logging in and found that our addictive technology facilitated the proliferation of false reporting, fomented white nationalist ire, and shattered the faith in discerning what is real and fake.

Certainly, the erosion Williams discussed in his talk had started well before Trump was sworn into office, but the reality TV star's victory was shocking enough to prompt many to take a closer look at the influence social media has had on their lives.

Other Silicon Valley technocrats echoed Williams' claims. In November 2017, venture capitalist and former Facebook executive Chamath Palihapitiya presented a talk at the Stanford Graduate School of Business. He had been invited to discuss how business can be conducted more ethically. But like Williams, he dedicated much of the talk to the problems of social media. 'The short-term, dopamine-driven feedback loops that we have created are destroying how society works,' Palihapitiya said. 'No civil discourse, no cooperation, misinformation, mistruth...' Asked how he felt about what social media has wrought, he responded, 'I feel tremendous guilt.'[28]

Following Palihapitiya's lead, Facebook co-founder Sean Parker addressed similar problems at an Axios event the same month. He criticized Facebook for trapping users in a 'social-validation feedback loop' and that he and other social media designers 'exploit[ed] a vulnerability in human psychology' – that is, the human desire to be social. The design feature that hooks users by meting out hits of dopamine is, Parker humbly bragged, 'exactly the kind of thing that a hacker like myself would come up with.'[29]

A year earlier, in a 2016 story in *The Atlantic* called 'The Binge Breaker,' journalist Bianca Bosker interviewed former Silicon Valley product philosopher Tristan Harris. In the piece, Harris admits that before joining Google he studied psychological manipulation as a Stanford graduate student in psychologist B.J. Fogg's Persuasive Technology Lab. After feeling guilty for addicting most of the world, Harris attends digital detox camps for ex-Valley employees and educates the public on the dangers of the attention economy.[30]

After earning millions helping to create some of the most

successful products in history, these turncoats have dedicated themselves to addiction prevention. Of them all, Harris is perhaps the most skeptical. He admits that attempting to stop Big Tech's single-minded push for more addictive technology seems futile. The persuasive tech industry is, as he says, a 'race to the bottom of the brain stem.'[31]

Nearly all of these tech leaders insist that the solution to the manifold problems of Big Tech is to design more ethical software. In doing this, they fall prey to an old belief system, one that privileges applied science and mechanics and compels individuals to worship at the altar of sublime technology.

The Digital Sublime

As a concept, the sublime has a long history dating back to Pseudo-Longinus in the first century.[32] It has been theorized in multiple disciplines over time, characterized by a few key tenets. Sublime things are astonishing, terrifying, powerful, and transcendent.[33] Mountains, for example, are sublime. They are colossal wonders of beauty and danger and represent the immeasurable greatness of the natural world. Mountains also evade conventional reasoning and description. Even trying to view one in its entirety with the limited perspective afforded to humans is impossible. They are weighty things that make humans feel small by comparison.

Throughout history, natural wonders were often the focus of sublime rhetoric, but great technological advances in the nineteenth and twentieth centuries changed this focus.[34] Because it seemed to lend awesome and unprecedented power to human ability, technology itself came to be seen as sublime. It wasn't difficult to consider the steam engine or the airplane sublime because they displayed tremendous power.[35] And because technologies are built by humans, then perhaps we too could wield great power over our lives, the environment, and each other.

Gradually, people believed in the sublime power of technology to shape the world. In *Channels of Desire*, Stuart Ewen writes about the public's response to the 'Corliss Engine' at the Philadelphia Centennial Exposition of 1876:

> The engine, displayed in a tabernacle to industry, gave middle-class onlookers a modern religious experience. Viewers of the engine were often reduced to tears in its awesome presence, swept away by the apparition that they were witness to the power of the universe, tamed and harnessed by reasonable, enlightened science. Separated from its industrial context, and enshrined within a house of secular worship, the engine stood as testimony to a world which could be controlled.[36]

Ewen situates the 'Corliss Engine' within a lineage of machines that were publicly displayed as 'examples of scientific development' and 'totems to the vision of a manageable universe.'[37] Above all, they represented the awesome power of progress.

Sublime technologies, as with anything sublime, are also terrifying.[38] Sublime things strike fear into people simply because they are powerful and awesome. And just as they open up possibilities for transcendence, they can also invoke evils.[39] With every new technology, no matter how sublime we may think it is, there arrives with it the possibility of it breaking down, erupting, or crashing. New accidents always follow new technological inventions. Their impressiveness is accompanied by the threat of instability or chaos.[40]

Communication theorist James W. Carey wrote extensively on a form of the technological sublime he called 'the electrical sublime,' which he claimed is a 'form of the industrial Edenic' rhetoric born from the Enlightenment project.[41] Carey traced the electrical sublime to the Industrial Revolution, which he considered the moment when the belief in progress as a kind of salvation reached a fever pitch.[42] Numerous thinkers throughout the twentieth century bandied about the promise that industry could – and *should* – solve the problems of humankind. Carey cited believers in the electrical sublime as Buckminster Fuller, the New Left, Zbigniew Brzezinski, and others who 'have cast themselves in the role of secular theologians composing theodicies for electricity and its technology progeny.'[43] These theologians believed technology is, above all, a gift and that this gift ought to be used for the greater good of human society.[44] Most importantly, Carey thought that the believers in the electrical sublime accepted as truth the ability for technology to reshape society and human beings. Future utopias could be realized thanks to technology's presumed unfettered power.[45]

Carey was critical of these beliefs. Whereas followers of the electrical sublime believed technology would usher in a freer, more democratic society and vanquish all centralized powers, Carey saw burgeoning technology as 'biased toward the recentralization of power in computer centers and energy grids, the Pentagon and NASA, General Electric and Commonwealth Edison.'[46] He understood what some saw as 'temporary manifestations of decentralization and democratization' to be nothing more than 'superimpositions upon a larger trend toward increased territorial expansion, spatial control, commercialism, and imperialism.'[47]

The imperial characteristics of the electrical sublime forged the bedrock of what sociologist Vincent Mosco calls the digital sublime. The digital sublime is seemingly unitary and democratic but ultimately as divisive as the electrical sublime. Mosco regards the information gap, or the 'digital divide' between 'haves and have nots,' as complementary to the 'digital divine,' or the apparent ability for technology to solve the world's problems.[48] Yet the so-called digital divine is a ruse. What initially appears as digitally sublime is in fact a more ruthless strain of capitalism that widens the gap between the poorer crowds and the rich few.

Mosco thinks the beliefs undergirding the digital sublime are nothing new. He sees the 'mythic promise of cyberspace' to be the same tune in a different key, a new spin on old faiths in pre-digital technologies like the telegraph and radio.[49] Although the digital sublime is a play on previous sublimes, it is also quite unique in our history. The wishes behind every manifestation of the sublime may be the same, but our current digital technologies are a result of the logic of twenty-first century post-Fordism and advances in communications technology. The 'new' media today – social media, smartphones, ubiquitous WI-FI, streaming, cloud technologies – could only exist because of specific technological breakthroughs and a specific kind of economic system that outsources labor. We can grant the digital sublime is like every other form of the sublime while also noting that our so-called sublime technologies of the current digital moment are quite unlike previous ones.

Those who believe in the digital sublime regard technology as the neutral tools we humans can use to build utopia. Today, Big Tech is the loudest evangelizer for digital utopia. To preach this gospel, former and current Valley technocrats lament the present state of software, which they view as unethical, and then propose solutions instructing the public that technology is the means by which we can form a better future. But they are not the only proselytizers. Western culture is teeming with stories, narratives, and media that all proclaim digital technology to be humankind's saving grace. Narrative media teaches us that existing and soon-to-exist digital technology will erase human suffering. Even those stories that critique the hold Big Tech has on our lives ignore larger structural problems or place the blame squarely on messy humans who just can't handle the awesome power gifted to us by the technocratic gods. If only we can learn to better use these tools to suit our needs, the story goes. But such tools are anything but neutral and should not be tasked with the responsibility of directing human existence. This faith in technology is an old myth, as Mosco writes, but it is spoken in new venues about new problems.

Chapter Two

The Race to the Bottom of the Brain Stem, or How Persuasive is Persuasive Technology?

If you own this child at an early age, you can own this child for years to come.
- Mike Searles, former president of Kids-R-Us[1]

The lyrics to Luke Bryan's 2017 smash single 'Light It Up' describe a fairly typical occurrence in the digital age. In the song, Bryan admits to constantly checking his smartphone in the hopes that the love interest he's pursuing will contact him. 'I wake up, I check it; I shower, and I check it,' Bryan sings. 'I get so neurotic about it.'[2]

Bryan, an award-winning country music entertainer whose albums have sold millions, is no amateur at writing songs designed to appeal to as many people as possible. His breakthrough hit, 2007's 'All My Friends Say,' is about waking up after a spell of hard partying and trying to remember the events of the night before. The song aided Bryan's rapid ascent and also helped to usher in a genre of mainstream country music that was more pop than anything that had come out of Nashville before. In an effort to remain relevant in a music culture bolstered by electronic pop and hip-hop, country pop – led in part by Bryan – continues to churn out offensively retrograde songs that appropriate elements from trap and hip-pop, depicting the rural South as a place solely useful to hold tailgates and portraying women as objects holding beers in pickup trucks. But Bryan, with his finger on the pulse of the culture, knew smartphones mediate our relationships in the early twenty-first century and addressed our contentious dependence on them in his otherwise vapid 2017 single.

'Light It Up' was a massive success because it reflects Western society's reliance on and obsession with smartphones. Nearly every facet of our lives is influenced by mobile computing, and the extent to which smartphones and other portable devices have profoundly affected the world is still not entirely known. We continue to live through the effects of this enormous technological paradigm shift, yet nearly everyone everywhere could attest to the power of inventions like the iPhone and social media like Facebook and Twitter.

Because we are so intimately linked with them, the technologies of our day interact with us psychologically in unique ways. They are designed to hold our attention, to manipulate neurochemicals and keep us tethered as part of a business model that many have started to question. As clinical psychologist Sherry Turkle explains, smartphones function as portable slot machines – games that go with us wherever we go and beg for our undivided attention.[3] Turkle writes that interacting with our technologies daily while living and working 'rewards [the body] with neurochemicals that induce a multitasking 'high.'[4] Like a gambling addict at the slots, we return to our devices over and over, thirsting for that high again.

In his essay 'A Nation of the Walking Dead,' Chris Hedges broadens the scope of our addiction to technology. Like Turkle, he also compares our 'personal computers' and 'hand-held devices' to slot machines, which 'draw us into an Alice-in-Wonderland rabbit hole.'[5] Slot machines and smartphones 'cater to the longing to flee from the oppressive world of dead-end jobs, crippling debt and social stagnation and a dysfunctional political system.'[6] Constantly checking our phone, like the character in the Luke Bryan song, feels good because it not only gives us a multitasker's high but also frees us, even for a moment, from a harsh reality, perhaps too soul-crushing to face directly. Hedges goes on to note that the gambling industry perfected the art of attention-stealing by targeting pleasure in a given gambler. Because player cards are monitored, casinos wait for the moment when gamblers stop playing to swoop in and seduce them with more tantalizing rewards. Tech algorithms function in a similar way. They manipulate behavior by keeping players in the zone, lifting them into a liminal space where they float among the lights and colors of the game.[7] The point isn't to win but, as Turkle writes, to 'stay in the game.'[8]

Silicon Valley's products are designed to keep us in the game. By assuming the human brain to be a machine itself, tech designers have created tremendously powerful devices that put the agency in the hands of algorithms. Unlike humans, these algorithms are inflexible. They cannot predict or quantify the mess of human life no matter what any Valley technocrat might espouse. They are, however, quite good at manipulating behavior, and in doing so, they simplify human life. When algorithms and bots determine human behavior, the upshot is a rigid society, each person in lockstep with the other, lines drawn sharply, creating binaries and eliminating measured debate. And with this erosion of social norms comes a plethora of mental health issues.

Persuasion, Sophists, and Coercion

Technology that changes behavior in people is often called persuasive tech. Stanford psychologist B.J. Fogg was the first to consider computers persuasive technologies, a field which he called captology.[9] According to Fogg's theories, persuasive technologies can be used to coach individuals towards better health outcomes, like quitting smoking or drinking more water. They can motivate people to live more sustainable lives by adopting environmentally-friendly habits. But they also pose numerous ethical quandaries. Who is at the other end of persuasive tech? When does persuasion become manipulation? Is persuasion always in the interest of those being persuaded?

Cultures throughout history have frequently regarded anything too persuasive as suspicious. Scholars believe the invention of persuasive arts can be dated back to the Sophists, who were itinerant teachers of rhetoric in ancient Greece and who often frustrated the philosophers of the day for their persuasive wordplay. Whereas Plato sought transcendent truths, teachers of rhetoric were either staunch pragmatists or eloquent connivers. To the ancient philosophers, the former were simply frustrating; the latter were suspect.

Gorgias, one of the most famous Sophists, spoke in a manner that amazed the audiences of ancient Greece. Listening to his words was like falling under a spell. Some were more easily lulled but others felt manipulated by this magical power. Somewhat of a proto-nihilist, Gorgias believed language is inherently deceptive and that no one can fully know anything with certainty. Because reality is structured by language, Gorgias believed nothing truly exists. In his infamous 'Encomium of Helen,' he offers a take on the myth of Helen of Troy, arguing that Helen was no match for Prince Paris' persuasive language, which came over her like a toxin. She shouldn't be blamed, Gorgias states, for language is so powerful that it can assault a person more violently than rape. The piece was controversial and continues to be to this day. Some audiences of his time were captivated by how he could change their minds using only his words. Others regarded the wordplay as sheer cookery.

The 'Encomium' was Gorgias' teaching demonstration. It served as a way for him to show prospective speakers what they could do if they took rhetoric lessons with him. A parlor trick, the 'Encomium' earned him students as well as a questionable reputation. At best, Gorgias was a trickster. At worst, he was a relativist who believed in nothing and sold his skills at manipulating opinions for a price.[10]

The *Dissoi Logoi*, an incomplete text by an unknown Sophist, demonstrates similar rhetorical maneuvers. The text makes the case that what is bad can be good depending on what perspective is taken. For example, it claims that 'illness is bad for the sick but good for the doctors. And death is bad for those who die, but good for the undertakers and the grave-diggers,' and so on. It was presumably an exercise to train orators to be knowledgeable about both sides in a debate, but like the 'Encomium' it seems to advocate for situational ethics and relativism, although scholars continue to debate this criticism.[11]

Ancient Greece was not alone in producing sophistic arts. In ancient China, the Mingjia practiced a form of sophistry not unlike that which was popular in ancient Greece. Classical Chinese ming bian, like Greek rhetoric, put an emphasis on persuasion, ethics, and reason, and some teachers distrusted the Mingjia's sophistic word games that argued right can be wrong and wrong can be right.[12]

We cannot be sure whether the Sophists or the Mingjia were relativists who used words to dazzle, confuse, or even manipulate others. We do know that some Greek philosophers of the time were highly critical of these persuaders. The Attic orator Isocrates famously condemned the Sophists and believed rhetoric should be used for the common good of society, not to stupefy like some linguistic sleight of hand. Instead of teaching them to search for universal truths as Plato did, or to play word games, Isocrates instructed his students to be model citizens and put their knowledge to practical use. He regarded the Sophists as scammers who made pedagogical promises they couldn't keep, like teaching students how to find happiness for a low price anyone could afford. Isocrates thought of rhetoric as an art and saw how the Sophists applied hard and fast rules to it in order to streamline and sell it like a product. They were untrustworthy in Isocrates' eyes, which meant they were unfit to teach things like truth and the common good. And they gave orators everywhere a bad name.

The common good, Isocrates believed, was not something

that generated from absolutes but from the everyday practice of doing good in public life. He was suspicious of grand, totalizing principles and favored local solutions. Each situation, he thought, carried with it its own principles that must be acknowledged to make moral decisions. If it didn't benefit the public good, Isocrates considered it unworthy of study. The Sophists, with their big promises and magical wordplay, were therefore damaging to the idea of a functional democracy.[13]

Sophistry is founded on the idea that language can change opinions and, by extension, people.[14] Even in ancient Greece, the fine line between persuasion and coercion was noted. Without moral teachers of rhetoric, persuasion could become a tool to manipulate others. In the hands of swindlers, words could sow discord.

Communication scholars Sonja Foss and Cindy Griffin have called persuasion inherently patriarchal because it is a practice of control and domination. And throughout history, platforms of persuasion have been helmed primarily by men, often with the effects of controlling the actions of everyone else. Even when less coercive, persuasion sometimes takes on a paternalistic flavor – the audience must be explained to and enlightened. Invented by men in ancient Greece and practiced by men atop podiums through time, persuasion has a long patriarchal history.

Rather than fostering relationships and promoting equality, persuasion has historically enabled competition, domination, and oppression. A kind of communication structured by 'equality, immanent value, and self-determination,' Foss and Griffin contend, is not persuasion but rather invitation.[15] Invitational rhetoric is inherently feminist, they claim, because it challenges the domineering nature of persuasion, making room for camaraderie over elitism. Its goal is 'the understanding and appreciation of another's perspective rather than the denigration of it simply because it is different from the rhetor's own.'[16]

If we agree that persuasion is a patriarchal art, then persuasive technologies are patriarchal inventions. An industry dominated by men, Big Tech may be the inevitable result of patriarchal persuasion. Of course, if we are to understand persuasion as inherently patriarchal, with feminist invitational rhetoric its corrective, we are also assuming essentialist definitions of gender. This makes little room for diversity, claiming all feminists are the same. And the binary of invitational and persuasive communication leaves out anyone not slotted into its grid.

Whether or not it's inherently patriarchal, persuasion can be used to force people to act in ways that might damage themselves or others. And it might prove financially beneficial for tech designers, who have built business models on persuasive tactics. The more technology interacts with humans, the more we come to depend on it. To sell products, software designers rely on a social influence model, which does business by making tiny suggestions that build up over time. Coupled with this is a peer-to-peer network that puts the thoughts and actions of others in front of users. Recommender systems suggest the content that they think users want in order to create a change – namely, buying something. These are all persuasive tactics meant to control consumers.

Persuasive technologies have come to be distrusted. They make promises they can't fulfill, often for free but really for the price of one's privacy and wellbeing. They precipitate artificial social networks, rigidifying the norms of communication and community and grouping users in echo chambers. It seems they are not practical for handling measured public debate, a hallmark of democracy and the common good. And they have a tendency to erode the mental health of their users.

A History of Illness

One of the earliest reports linking mental health to social media usage was published in 2013 by Ethan Kross and his team of psychologists at the University of Michigan. At the time, 500 million people used the social media service, but no study had been done on its mental health effects.[17] The particularly shocking part of the study wasn't that Facebook could make a person's life worse. Instead, the findings indicated no other factors tempered the ability for Facebook to influence or disrupt one's life. 'On the surface, Facebook provides an invaluable resource for fulfilling the basic human need for social connection,' writes Kross and his team. 'Rather than enhancing well-being, however, these findings suggest that Facebook may undermine it.'[18] Put simply, Facebook does not exacerbate depression or loneliness, according to Kross. It *causes* them.

There had been damning reports about the impact of new communications technologies on mental health before the Kross report, but those largely concerned the mental wellbeing of workers in other countries who manufactured iPhones and other consumer electronics. Three years before Kross and his team threw light on Facebook's psychological effects, several employees at Foxconn in China committed suicide by jumping from factory buildings.[19] A major manufacturer of iPhones for Apple and other electronics for Hewlett-Packard and Sony, Foxconn Technology has a history of questionable labor practices leading to employee mental health crises.[20]

The Foxconn suicides were often chalked up to the high suicide rate in China, but they also served as premonitions of events to come. During his talk at the Stanford Graduate School of Business in 2017, Chamath Palihapitiya noted that 'bad actors can now manipulate large swaths of people to do anything [they] want.'[21] Palihapitiya, of course, is not referring to the bad actors like Apple and Foxconn that pushed workers to suicide in 2010. But he is right that social media and digital technologies give people the power to control others.

We Machines

Because Valley technocrats believe the brain to be like a

machine, they think they can hack it. And because they have had tremendous success at manipulating both society and psychology, they also believe they have the authority to propose solutions to mental health problems.

There is a long history of the brain being likened to a machine, and the metaphor is commonly used in lay discourse. We talk of people short-circuiting, neural networks being wired in specific ways, and brains resembling computer chips.[22] Hacking the machine brain allows technocrats the ability to manipulate dopamine and serotonin with design features. By simplifying the mind with machinic terms, the technocrats then put forth a basic process of addiction that is as linear as pulling levers and programming people.

Katie Lynn Walkup and Peter Cannon, researchers at the University of South Florida, have noted the danger in reducing addiction to such linear, simplistic conceptions. They claim that 'addiction has multiple ontologies,' such as 'social factors and poor health care,' which clinicians and physicians sometimes ignore.[23] Positing singular causes for mental illnesses could result in clinicians '[e]mphasizing one conception of addiction over another,' thus leading patients to 'believe that addiction is only caused by one factor.'[24] By ignoring social contexts, some mental health workers perpetuate a diminished explanation of mental illness.

Focusing attention on addiction itself also ignores a multitude of 'underlying mental health concerns that may contribute to addiction.'[25] It's a nice fantasy to think addiction occurs in a vacuum, that alcoholics and patients addicted to opioids have chosen to use and abuse substances as rational, logical actors stripped of context. This fantasy leads many to criticize addicts in rote ways. *Why don't they quit? Just say no.* Similarly, when Valley technocrats aren't proposing tech fixes to their own tech problems, they ignore context by telling users to just log off and escape the addiction.

Other solutions proposed by Valley turncoats involve practicing some form of pop mindfulness, a process made popular in the West by the self-help industry, which traffics in selling mind cures to an anxious and depressed neoliberal society. According to communication scientists Lucas Youngvorst and Susanne Jones, there are typically two components involved in achieving mindfulness. First, a person must focus thoughts on the present.[26] By remaining in the present moment, the mindful individual stays centered and calm.[27] Mindfulness practices emphasize the fact that the past and future cannot be controlled. All that matters is the here and now, the present moment you are in.

Mindfulness also involves 'emotional detachment' and encourages practicing individuals to refrain from reacting to or judging what occurs around them.[28] Those in mindfulness states are merely present observers aware of situations and experiences around them. The goal is to think 'without having preconceived notions about what ought to be thought, felt, and done.'[29] This is achieved by encouraging patients to recognize their attachment to 'ideas, expectations, or rules' and then guiding them to disconnect from those attachments. Though mindfulness is taken up by the self-help industry and often sold through adult coloring books and digital detox camps, it can be a beneficial strategy employed by practicing mental health clinicians for some client-patients seeking treatment. For example, mindfulness is often correlated with empathy and can help patients manage complicated or distressing emotions.[30]

But mindfulness is not safe from the roving eye of capitalism, which seeks to commodify everything, including mental health.[31] Capitalism does this successfully by posing problems that only it can solve. In *Captains of Consciousness*, Stuart Ewen provides a robust history and thorough critique of capitalism's creation of advertising and the tactics used to ensure people keep buying things. Ewen writes that advertising under capitalism identifies and exploits 'feelings of social insecurity' to 'habituate men and women to consumptive life.'[32] Like Facebook co-founder Sean Parker, who also confessed to exploiting the human need for social connection to 'sell' social media, advertising executives understood early the power in targeting insecurity and in posing purchasable solutions to problems. From its beginnings, advertising told educated consumers that *they* were the problems to be solved and induced in them a 'critical self-consciousness in tune with the 'solutions' of the marketplace.'[33] It is not enough for companies to 'argue for products on their own merit.'[34] Ideas and desires, although not necessarily created *ex nihilo*, have to be channeled and marketed to the critical individual to increase profit.[35]

This does not mean advertising injects ideology into all of us and turns us mindless as a result. Because a great deal of the West enjoys consuming, capitalism is there to offer products on tap. But capitalism also knows it is in the business of selling solutions to problems it frames. Those solutions, we often believe, will deliver us from suffering and, in the twenty-first century, are increasingly technological.

Big Tech's menu of inventions includes devices designed to solve problems that perhaps were never considered to be problems in the first place. One such human problem is the state of not knowing something. If the smartphone is an unprecedented invention in human history, it is simply because it allows the bulk of knowledge to exist at our fingertips. Just as no one may ever get lost in the digital age, little is left unknown as well, and mobile computing devices, while providing ceaseless entertainment and distraction, are gateways to information bliss. It would seem nothing is greater than our addiction to information, for the more we know, the freer we feel.

While smart devices soothe the pain of loneliness by connecting users in social networks, wearable technology that monitors daily physical activity can reconnect wearers to something they seemed to have lost in the hyperspeed shuffle of the digital age: their bodies. Activity trackers provide constant, quantified information where previously there was little. By medicalizing daily routine, these fitness trackers give users a ceaseless stream of information about the human body that offers freedom from the panic of not knowing something.

Counting just how many steps are taken in a day leads wearers to believe they are participating in and strengthening their health. But the reality is much different. Wearers assume fitness trackers are medical devices, but they function more as surveillance. They also accrue a massive amount of information, supposedly for the user's empowerment, to quantify the self in increasingly precise ways. This leads to questionable medical benefits and increased anxiety. Wearing a tracker fails to free the consumer from risk.

Trackers and biosensing wearables are examples of digital technologies that are marketed as exits from material flesh into the utopia of pure, clean information. It is a fetish for people who are obsessed with constant improvement and growth. Whether these technologies remain or are replaced by others, the corporate wish behind them will continue. That wish is for sublime and awesome technology to solve all our problems.

Big Tech's Colonization of Education

Big Tech has invaded nearly every area of modern life, from health to politics, and not even the education system is safe from its tendrils. In schools across the country, educational apps and services like ClassDojo, Schoology, Otus, D2L, Google Classroom, and Canvas are employed to create blended learning environments for students. They purport to streamline assignments and keep parents updated on their children's progress. Teachers can use them to collect homework, keep attendance, facilitate classroom collaboration, administer tests and quizzes, share files, schedule exams, and do anything else a teacher might need. Although they perform these functions and sometimes make life easier for teachers, they also harvest data from young people, a demographic highly sought by advertisers.

Schools are increasingly becoming sites of ad penetration. Since at least the end of the twentieth century, corporations have been targeting students in their own learning environments in order to better sell their products. What were once ad-free zones are now epicenters of marketing, in which students are held captive while ads zero in. Scoreboards, buses, and lunchrooms are plastered with ads, while schools are offered incentives to take field trips to nearby corporations. Soda companies and automobile manufacturers sponsor schools. And tech companies are the latest to colonize learning spaces, providing educational apps in exchange for access to a valuable age group.

One such app is the popular and controversial ClassDojo, which takes the social network model and applies it to the classroom.[36] After registering for the app, teachers can create class profiles, post updates for parents to see, and track student progress. If a student acts out or breaks a rule, the teacher can punish her by docking points from her profile.[37] Although it claims to make classroom management easier, ClassDojo inevitably functions as a way to quantify and control student behavior.

Educational apps promise to help overworked and underpaid teachers by quantifying the behavior of students and keeping parents in the know, but this is merely an excuse for corporations to maintain their presence in the classroom. And students are the ones most negatively affected. ClassDojo teaches young people to understand digital technology as a natural part of life, and it poses a risk to those who might live in hostile households, whose only refuge from abusive parents might be the classroom. It also normalizes surveillance and competition and fuels anxiety.

Going Back to School

As schools are targeted by Big Tech and transformed into miniature crucibles of control, some young people yearn for lives without the stress and distraction of digital devices. They grow nostalgic for older education systems and turn to popular media representations of pre-digital schools to assuage their nostalgia. This is not a unique phenomenon, as nostalgia appears everywhere throughout Western culture.[38]

So many today are nostalgic for adolescence and childhood – perhaps not the childhood they lived through but one free from the tyranny of digital tech. They watch series like *Stranger Things* and *Everything Sucks!* to get a nostalgia fix and gaze longingly at a world in which technology either makes sense or doesn't destroy our lives. These kinds of popular media present adolescence as a halcyon period of freedom. Characters in these stories deal with the usual trials of awkward adolescence and school daze, but they do so without educational apps, data rooms, biometric locks, or closed-circuit television. They attend fantasy schools, where daily concerns are manageable and controllable, and live freely without the terrors of the twenty-first century.

The more complex and technologized our control society becomes, the larger the nostalgia industry will grow. As social problems are solved by increased monitoring and corporations uphold competition as the means to achieve personal fulfillment, the nostalgia industry will continue to churn out representations of pre-9/11 pop culture, making a killing in the process. Trapped in their schools, quantified and controlled, young people are no match for the neoliberal logic of Big Tech.

The concerns about teen health and smartphone usage have been widely reported. Psychologist Jean Twenge, in particular, has dedicated much of her work to understanding just how severely digital technology damages the mental health of young people. She's found that 'post-Millennials' suffer from mental illnesses at an unprecedented rate. They're going out, having sex, driving around, and dating far less than previous generations of young people. This dramatic shift occurred sometime between the end of the 2000s and the second half of the 2010s, a turn she attributes to the rise of smartphones. Instead of doing what teens have always done, they're quartering themselves in their rooms and anxiously scrolling on their phones – all day, every day.[39]

It has always been easy to criticize the habits of young people, who, according to some conservative adults, should be reading the great classics of literature instead of racing hot rods and listening to Elvis, or posting to social media and texting. Nothing comes across quite as pompous as an older generation decrying the younger ones for doing things differently than it did. It might seem Twenge is guilty of this, but a few factors differentiate old-fashioned hand-wringing from genuine concern about ubiquitous digital technologies.

Yes, the health of young people is at risk when Big Tech colonizes learning spaces, but so is their privacy. Educational apps collect an enormous amount of student data, often without their consent, and with no way to opt out. This leads to staggering privacy concerns, as a company like Google mines personal information for advertisers to peruse. Slowly, thanks to these giant tech corporations, the right to ambient privacy is eroding – a process that now starts in the classroom.

And young people are quite aware of the problems of persuasive tech. Much of this awareness is due to the wide coverage of tech addiction in the popular press and the ease with which that information can be accessed. If teens are always on their phones, like many of us are, chances are they've run across an article, meme, or post commenting on the effects of social media. Their awareness is also due to a shift in how we talk about mental health, which has slowly become less stigmatized in the Western world, although there is more de-stigmatizing to be done. But much of the current culture still doesn't know how to pose reasonable solutions to the problems of runaway persuasive technologies. Currently, two popular options are frequently offered: either we allow the technocratic turncoats to invent more ethical software and risk trading the devil we know for one we don't, or we run headlong into an idyllic past.

Stressed Out

Towards the end of 2015, the musical duo Twenty One Pilots released their single 'Stressed Out,' a massively successful song about growing up in a time of harsh social instability. Taken from their second major label album, *Blurryface*, the song shot up the *Billboard* Hot 100 chart, cresting at number two, and earned a Grammy for Best Pop Duo/Group Performance.[40] In 2019, it was certified 8x Multi-Platinum in the U.S.[41]

The song is simple. In it, the duo sings about the anxiety of influence and the pressure to come up with something truly new. 'I wish I found some better sounds no one's ever heard,' vocalist Tyler Joseph sings. 'I wish I found some chords in an order that is new.' But then Joseph quickly pivots in the first verse to address a more fundamental problem. 'I was told when I get older, all my fears would shrink,/ But now I'm insecure, and I care what people think.'[42]

When the chorus arrives, Joseph is even more direct. 'Wish we could turn back time/ To the good old days/ When our Momma sang us to sleep but now we're stressed out.' This is a bold confession for a pop act to make. Here is a world-famous band penning a song about the desire to be a kid again – a wish that struck a chord with people across the world. But it ultimately makes sense given the context. 'Stressed Out' is a reflection of our age of nostalgia, the kind that yearns for both a previous society *and* the carefree Eden of childhood, when 'nothing really mattered' as Joseph sings. The longing for childhood is old as humans, but in the early years of our century, it has material aims, which the song addresses. In a particularly striking instance, Joseph sings that he would rather play in tree houses than bear the burden of student loans. For Twenty One Pilots, and for many in the U.S., turning back the time not only means becoming a kid again but also living in an historical period before college tuition skyrocketed.

'Stressed Out' is a deeply sad song because it openly addresses the reality that so many in the West (particularly in the U.S., where student loan debt has reached crisis levels) would rather revert to infancy than continue living in a brutal society – one where everything has been done and nothing is new, where social media's cult of competition trains us to compare ourselves with others, and where, from an early age, the impulse to ruthlessly accrue wealth is driven into our thoughts like a rail spike. At the same time, it's a hopeless song because the solution it presents is to retreat into the false utopia of childhood. It misremembers the past as a wonderland where possibilities were endless and stability wasn't in short supply. This starry-eyed view of the past, comprised of our personal histories and larger historical narratives, is toothless and safe. It is a reactionary wish because it refuses to recognize the past for what it was: a time with its own cadre of problems from which we can still learn. We can gaze into the past and see that, for example, college tuition was lower, administrators weren't running higher education, corporations hadn't yet colonized classrooms, and debt wasn't suffocating. Knowing these facts, we can begin to imagine alternatives to our current social fabric. But until then, yearning for mother and the supposed freedom of childhood allows larger structures of power to sustain.

It is no coincidence that the mental health crises of our time exploded in the wake of Big Tech's rise. Our current technologies were designed to hook users, and they now have infiltrated every part of modern life. They are indispensable for many people and help to lubricate free-market capitalism. At the same time, they carry profound mental health risks. They connect people from all over the world, but they continue to atomize an already lonely society. They provide us constant information, keeping us in the *know* and in the *now*, and as a result, a nostalgia industry has scaled to sell products providing respite from a complex world.

Sherry Turkle makes clear that what we are addicted to aren't the technologies themselves but 'the habits of mind that technology allows us to practice.'[43] These habits are numerous and all too human: multi-tasking, speed, distance, and the desire for control. But it is becoming obvious that we cannot live as an orderly society this way. The result of a networked neoliberal order has been disastrous, and one of the more destructive effects has been an increased presence of nostalgia – the kind that tricks you into thinking *everything* in the past was better, or it must have been because the present is so unbearable. From the desperate cry of 'Make American Great Again' to the simulated retro worlds appearing in mainstream television and cinema, the West dreams of the past.

Chapter Three

No More Futures: Pre-Recession Nostalgia in the West

Nostalgia ensures that certain things stay before us: the things we miss.
- *Sherry Turkle*[1]

Dig up the past, all you get is dirty.
- *from* Minority Report *(2002)*[2]

After releasing her chart-topping studio album *Red* in 2012, Taylor Swift found herself on a beach with musician Jack Antonoff and his then-partner Lena Dunham.[3] Antonoff, the frontman of retro pop band Bleachers and former member of the indie crossover act Fun, had met Swift at the MTV European Music Awards in November 2012, and the two developed a friendship over the course of the year.[4] But it was on a vacation together that the pair began laying the plans for Swift's fifth studio album, one that would ultimately draw more inspiration from mainstream pop music and less from the country sound that made her famous. They bonded over a mutual love of 1980s pop acts like Fine Young Cannibals and Yazoo and eventually decided to work together.[5] With Antonoff's help, Swift signaled that her next album would be a major stylistic change, with *Red* serving as the transition from her confessional country sound to pure stadium-filling electronic pop.

Whereas *Red* only flirted with pop, Swift's fifth album was a full sonic pivot for the twenty-four-year-old artist. She had wanted to cross over from country to pop for years and, along with her team, had seen the writing on the wall when the more pop-friendly songs on *Red* exploded onto the charts.[6] There was, of course, hesitation and outright resistance among some in her inner circle.[7] Why would the biggest country star on the planet seek new markets? Why go pop? The reality was that Swift had always been pop, and she knew it. She also knew the country music industry had been very successful relying on pop production for years. So, Swift dropped the fiddles and steel guitars and drew instead from a retro aesthetic that had sustained wide appeal in indie music since the mid-2000s.

In August 2014, Swift released the cover art for her fifth album, *1989*, and it was clear Antonoff's love of 80s aesthetics would be palpable on the record. The cover shows a faded instant photograph of Swift wearing a long-sleeve shirt with seagulls on it and 'T.S. 1989' scrawled in black marker at the bottom of the white frame. Although the album's first single, 'Shake It Off,' bore the mark of contemporary pop mastermind Max Martin, the songwriter behind most pop number-one hits since the mid-90s, *1989*'s aesthetic was noticeably retro – a blend of Antonoff's cinematic indie pop and 80s electro.

Indie Nostalgia

Much of mainstream music's obsession with the sounds of the past can be traced to *Funeral*, the 2004 debut album by Canadian rock band Arcade Fire. A massive critical success, *Funeral* featured pastoral instruments like the accordion, harp, and hurdy-gurdy and, with its rustic album art and lo-fi production, signaled a back-to-roots aesthetic for popular music. While indie rock acts like Fleet Foxes and Beirut followed in Arcade Fire's footsteps, incorporating folk and appropriating traditional music from outside the West, other indie pop groups drew heavily from the sounds of 1980s Western popular music. By the release of *Red* in 2012, the indie music landscape was littered with retro pop influencers and also-rans: Neon Indian, Washed Out, Toro y Moi, Memory Tapes, and M83, whose 2011 smash single 'Midnight City' blew clear the path for pop artists seeking to capitalize on the cultural thirst for the 80s.

'Midnight City' was a watershed moment for indie pop that helped propel retro into the mainstream. It is the sound of 80s electro filtered through the hi-fi production quality of twenty-first century synthpop. This kind of nostalgia that blends old and new – often known as retrofuturism – had been circulating in culture for a few years prior to the release of 'Midnight City,' but the single, and M83's 2011 opus *Hurry Up, We're Dreaming*, proved 80s nostalgia could be commodified and sold widely.

Three years after 'Midnight City,' Swift's *1989* cemented the nostalgic monoculture, and the aesthetic trends of some online music genres suddenly became mainstream. The third single from *1989*, 'Style,' has obvious roots in synthwave, an Internet-born electronic music genre that appropriates sonic and visual elements from 1980s video games, slasher films, science fiction, and teen movies.[8] Characterized by arpeggiated synthesizers and analog production, synthwave harks back to the soundtracks of Vangelis, the synth scores of John Carpenter, and the *Out Run* video game, as well as the music of German electronic group Tangerine Dream.[9] The genre is a retrofuturistic attempt to capture the sounds of 1980s American culture and present them as unadulterated as possible in a twenty-first century context. Producers typically refrain from sampling directly from 1980s music but instead create new compositions that mimic the sounds from that decade.

Stockholm-based synthwave outfit Forêt de Vin describe their music as 'the songs the record company forgot to release 30 years ago,' which is also an apt description of most synthwave music to date.[10] Artists like The Midnight, Dynatron, and FM-84 traffic in 80s pop re-creations that often sound more like a simulation of the 1980s than the decade itself. The aesthetic markers of the time are present (analog synths, gated drums, images of muscle cars and gleaming cityscapes) but the quality is noticeably boosted. Synthwave is the sound of the past filtered through the present, an analog dream in a digital age. And *1989* marked the moment when that dream became a cultural reality.

Synthwave producers are influenced as much by M83 as *Drive*, the 2011 Nicolas Winding Refn film about a getaway driver who finds himself embroiled in a criminal situation. The soundtrack features synthwave pioneers College and Kavinsky, and even the font for the film's title is unabashedly retro: a pink cursive scrawl not unlike the text seen on film posters for *Mannequin* (1987) and *Sixteen Candles* (1984). *Drive*'s extended sequences of night driving accompanied by hi-fi analog synths inspired a generation of synthwave producers.

Both *Drive* and 'Midnight City' facilitated a mainstream acceptance of 80s retromania that would prime audiences for Swift's *1989*. But when M83 decided to mine the far-reaching corners of 80s nostalgia on *Junk*, his 2015 follow-up to *Hurry Up, We're Dreaming*, the reaction was less enthusiastic. In the four years between the albums, nostalgia fixed itself in the monoculture and took on a particular form: it referenced previous decades but did so without losing its footing in the present. In other words, nostalgic media in the early 2010s dipped deeply into the past but not so deep as to alienate present consumers. *Junk*, on the other hand, is nostalgia approaching the uncanny valley, an attempt to force dated sounds into the present without distance or revision. Listening to it is a strange experience. The album digs up the chintziest, cheesiest sounds of 80s music without filtering them through the present. In doing this, M83 commits an historical violence by colliding the pure past, warts and all, with the present. Much of the music press picked up on this when *Junk* was released; the negative reviews were quite unlike the praise for M83's previous records.[11]

Synthwave has spawned several of its own subgenres, some playful and others frightening.[12] One example of the latter is fashwave, short for 'fascist wave.' Like synthwave, fashwave refashions the sounds of 1980s video game music and synth scores but pairs them with audio samples of anti-Semitic slurs, white nationalist commentary, and Trump speeches (Trumpwave is itself a subgenre of fashwave).[13] Artists like CYBERNΔZI, Xurious, and Behemoth include Nazi imagery, Trump memes, and clips from the 80s in their music videos. Because it blatantly displays white nationalist and authoritarian imagery, fashwave sparked a moral panic when it bubbled up from the depths of the Internet.[14] White supremacist Richard B. Spencer openly endorsed the subgenre, and Santa Fe High School shooting suspect Dimitrios Pagourtzis posted fashwave-related iconography on his now-deleted Facebook page, which kickstarted a fear that fashwave might indoctrinate people to commit hate crimes.[15]

Charting the recent music trends in Western mainstream media is one way to study the increasing popularity of nostalgia, but there are other paths one could take. Movies, video games, streaming series, and advertising rely heavily on nostalgia for a very particular point in history.

What has transpired over the course of the present century is a collective re-envisioning of the immediate past. Western culture has been plundering the 80s, 90s, and early 2000s for media content to satisfy the appetite of audiences in the digital age. There are several reasons why so many are nostalgic for those particular periods. The terrorist attacks that opened the century on September 11, 2001, the financial catastrophe of 2008, and the invention of social media and smartphones have all contributed to increased levels of cultural nostalgia. Rattled by destructive global events and armed with the most powerful communication tools ever invented, the West has plunged into its own past, committed to recapturing what might have been lost with the dawning of an eternal present, when daily political traumas and digital sirens shackle us to *now*.

Nostalgic Dreams

Forty-seven remakes, reboots, or sequels were released in American theaters in 2017.[16] Of these films, *Star Wars: The Last Jedi*, an entry in the wildly successful *Star Wars* franchise, was the highest-grossing release of the year, breaking one billion dollars worldwide, and *It*, an adaptation of the Stephen King novel set in the late-1980s, became the highest-grossing horror film of all time.[17] In fact, the top ten highest-grossing films of 2017 were sequels, remakes, or installments in extended universes.[18]

Based on these figures, it seems nostalgia sold very well at the box office in 2017, yet nostalgia has been circulating widely in American popular culture for many years. Television reboots, throwback mainstream pop, and nostalgic re-imaginings are only a handful of recent entertainment media remaking, revising, or expanding the sights and sounds of former decades. Nostalgia also appears in advertising campaigns, on book covers, and even on soft drink labels. In the twenty-first century, content creators, both mainstream and independent, mine the past for creative ideas worth bringing into the present, and the upshot is an American culture with its gaze fixed on the good old days.

Since the Great Recession, nostalgia has become nearly ubiquitous in American culture – so much so that it's been employed on the campaign trail as well. During the run-up to the 2016 presidential election, republican candidate Donald Trump pledged to 'Make America Great Again,' a nostalgic appeal that ultimately served him well enough to secure a victory over former Secretary of State and democratic hopeful Hillary Clinton.[19]

After the election, commentators attempted to explain Trump's nostalgia for a return to a former iteration of American empire – one that, some believed, would usher in a new era of intolerance and violence.[20] Drawing a line from the 1964 Republican presidential nominee Barry Goldwater to Ronald Reagan and on to Trump, rhetorical scholar Mary Stuckey notes that 'a specifically racialized version of...politics' that, ultimately, '[does] not include equality' shows up time and again throughout American history.[21] She argues the intolerance circulating in previous conservative rhetoric was 'unleashed, if not unhinged, by...the Trump campaign's bid to 'Make America Great Again'': a campaign which flooded the country with the 'ugly' rhetoric of 'white nationalism, gender bias, and other antidemocratic elements of the polity.'[22]

We do not yet know the implications of living in a country that continues, as Fredric Jameson writes, to 'imitate dead styles.'[23] Although researchers have tried to outline theories of nostalgia over the years, and some have even focused on pop culture's recycling tendencies, nostalgia in our time is particularly important to study because it might shed light on the many problems plaguing our democracy, such as political division and ideological tribalism.[24] We can also hope to understand nostalgia not only as an individual or even communal emotion that often promotes prosocial behavior but also as a rhetorical tool that can be utilized towards dangerous ends.

The Presidential Election of Donald Trump

Donald Trump's rise to presidential prominence further divided an intensely partisan public in the United States. His announcement of his candidacy given at Trump Tower on June 16, 2015, included several positions upon which he would run his campaign. He declared that he would build a wall between Mexico and the United States; the Affordable Care Act was a 'disaster called the Big Lie;' he would be 'the greatest jobs president that God ever created;' he would refuse donor financing because he is 'really rich;' and the 'American dream is dead.'[25] It was during this speech that Trump trademarked his promise to 'Make America Great Again,' a vow that had circulated in American presidential rhetoric for decades. Though Trump claimed to have invented the slogan, both Ronald Reagan and

Bill Clinton employed the phrase on their respective campaign trails.[26] And like them, Trump discovered the phrase would ultimately benefit him.

Trump represented in the minds of many Americans an emblem of the fringe right and the manifold forms of intolerance the group champions.[27] The heated election that resulted in a surprise win for Trump inspired a tidal wave of social movement protests.[28] The 2017 Women's March, which currently holds the record for the 'largest day of protests in US history,' occurred the day after his inauguration, and numerous other resistance movements, both in the streets and on social media, coalesced in solidarity against Trump and the wider rhetoric of hatemongering he endorsed.[29]

At the same time, rightwing rhetoric from the center-right to the most fringe corners of the alt-right positioned Trump protesters as 'social-justice warriors' and 'snowflakes,' among other derogatory terms.[30] These insults existed in some form before Trump's election, but they gained greater rhetorical currency after his victory and are often trotted out to smear social movements that seek to resist a culture of hostility. What rightwing pundits, commentators, talking heads, politicians, and publics saw in the protests against Trump were radical leftists bent on wielding identity politics as a dangerous weapon. Against this backdrop of intense partisanship and heated culture wars, Trump was glorified as the strong leader for which some groups in the country had been yearning.

Trump also achieved mainstream legitimacy in an era saturated with nostalgia, and he knew how to work the emotion to win votes. The rallying cry of 'Make America Great Again' cued certain populations to believe he could feed the present back into the past and re-establish a pre-9/11 social reality. During a time when so many were desperate for exit from the present, he weaponized this universal, controversial emotion that dates back centuries but had become a primary tool for directing politics in the twenty-first century.

Pre-Recession Nostalgia

Scholars in multiple disciplines have endeavored to come to terms with nostalgia. The word itself first appeared in 1688 in the dissertation of Swiss doctor Johannes Hofer.[31] Believing in a cure for such a condition, Swiss physicians of that time sent their nostalgic patients on trips to the Alps and prescribed opium and leeches as pre-modern remedies.[32] Over time, the term gradually shifted from a medical illness to a cultural malaise, a change often attributed to industrialization.[33] Some categorize nostalgia as an emotion; others, a rhetorical invention. It is often written about as both a personal and collective feeling. There are many competing claims, and consensus is rare.

Social scientists and psychologists continue to grapple with understanding what nostalgia is, how often we respond with nostalgia, and if nostalgia is healthy. Psychologists Ulrich Orth and Steffi Gal consider it 'an emotional experience' that 'involves...memory' and is characterized by a 'wistful mood' and 'sentimental yearning.'[34] Evidence shows some people are more prone to nostalgic feelings than others.[35] And nostalgia differs from remembrance, though remembrance factors into the emotional process that yields nostalgia. Essentially, as psychologist Krystine Batcho writes, 'one can remember without being nostalgic, but one cannot be nostalgic without remembering.'[36]

Nostalgic rhetoric peaks during times of crisis as individuals seek meaning and stability.[37] Literary scholar Elena Oliete-Aldea considers the financial meltdown of 2008 a catastrophe that shattered the 'optimistic belief of a linear progression of history for the better' and ultimately gave rise to a peak nostalgic period.[38] 'Against this background,' Oliete-Aldea writes, 'the past becomes the only stable point of reference that enables one to safely cling onto a secure cultural identity.'[39]

Yet the crash of 2008 did not singlehandedly usher in this crisis period of fragmentation and eventual nostalgia-baiting. Instead it functioned as one crisis in a process that also includes the 'liberalization of global economies' and the 'spiral of global terror and fear' brought on by the terrorist attacks on September 11th.[40] Oliete-Aldea notes that prior to 2008, the conversion of the world into a 'borderless space' increased the anxiety of an already fragmented American culture.[41] She makes clear that the 2008 financial crisis only exacerbated this anxiety of a fragmented world in which the codes of meaning are scrambled beyond recognition.

Struggling to find stability, individuals living through a global crisis look to 'mainstream cultural products' like Hollywood films, television, popular music, and online media.[42] 'With no faith in the future,' writes Oliete-Aldea, 'what remains is to hark back to past times where socio-economic relations, if not perfect, were, at least, better defined.'[43] This conscious act of looking backwards – what Roland Robertson calls 'wilful [sic] nostalgia' – occurs at historical junctures when societies become unmoored.[44] National identity is threatened, and traditions are invented to soothe the anxiety of meaninglessness.[45]

Pre-Recession nostalgia is a particular strain of wilful nostalgia that recycles mass media from the years leading up to the first decade of the twenty-first century in order to present a simplistic version of history. Unlike other conceptions of nostalgia, pre-Recession nostalgia is most productively understood as both public *and* private, historical *and* personal. The time period which it references is one before the Global War on Terrorism, Web 2.0, Big Tech, and of course the financial crash of 2008. It is an older time when, to quote nostalgia scholars Doane and Hodges, 'men were men, women were women, and reality was real.'[46]

Pre-Recession nostalgia is also presentist because it privileges contemporary ideas over an actual 'past that was different,' as media theorist Ryan Lizardi claims.[47] A presentist version of history is one accented with anachronisms that nostalgic individuals invent as they misremember the past. In presentist discourse, the past is distorted to appear more appealing than the present. This move is inevitably dangerous; the past makes promises that history cannot keep. For some, confusion eventually sets in: what is real and what am I imagining?[48]

For others, there is no confusion. There is only the presentist version of the past yearned for in this moment of global crisis. Presentism functions as the primary perspective of the tourist, who views the world from a distance and stares only at what manifests on the surface.[49] Cultural critic Marita Sturken has written extensively on this 'tourism of history,' which she defines as 'the mediating forms through which the American public is encouraged to experience itself and the nation's relationship to global politics and world history through consumerism, media images, souvenirs, popular culture, and museum and architectural reenactments.'[50] The tourist watches history from a removed perspective, couched in the comfort such a distance affords, and typically surfs and skims uncritically. And since tourists do not critically engage with what they watch, according to Sturken, they operate in an atomized mode, separated from the watched object or event as well as the other tourists who may also be watching. Detached from any historical referents and suffering from the amnesia of a fragmented present, the tourist injects present biases into the past.

9/11 and Futurelessness

In the wake of 9/11, many looked fondly to the past, a tendency that has not let up since. In fact, it has remained for well over a decade thanks in part to the digital technologies we use daily. In a time of increased technological reliance, when much of our lives are structured by digital tools, many want respite from a hyperspeed world. At the same time, the very structure of the Internet allows all points of history to be visible at once, at any time, with just a click or swipe. When anyone with a device or computer has unfettered access to all of history, the result is a penchant to dig up the past. This isn't a particularly novel impulse in Western popular culture, but the Internet does make it easier to surf through historical periods and mine the popular tropes from any time period.

But it isn't just any period that circulates widely in Western culture. Presently, pop culture fixes its gaze on the decades leading up to September 11th, the Recession, and the invention of social media – a period of tremendous capitalist expansion and social upheaval, when pop culture fought hard to colonize consumers' thoughts by launching extensive advertising campaigns. What percolates into the present aren't the more radical elements from the late twentieth century but far more reactionary narratives. In this way, pre-Recession nostalgia reinforces the beliefs among contemporary publics that in the past the future was hopeful, but that in the present America suffers from what Robert Jay Lifton calls 'futurelessness,' which is a hopeless collective feeling that no future will transpire.

This national nostalgia for a time before the twenty-first century is a step in what Robert Jay Lifton calls the 'stages of response,' which trigger during the aftermath of a harrowing national catastrophe. Immediately after 9/11, Lifton writes, America was reeling from the 'shock of sudden vulnerability.'[51] Unprepared for such a violent event, the public experienced a 'sudden shift in [the] perception of the American landscape.'[52] The result was

> pervasive, timeless dread, the feeling that we can never be free of this ominous threat, that we may be awaiting a moment that could dwarf 9/11. It has become increasingly difficult to envision a future free of that dread. This sense of *futurelessness*…has to do with a break in the flow of generations, an interruption of human continuity.[53]

Lifton's 'stages of response' chart the process by which futurelessness is employed as a reaction to national crises. As futurelessness increases among the public, many start to have 'doubts and resentments about the directions our leaders' mission has taken us in,' and some look to strong leaders to protect them from further shock, which, according to Lifton, is the final stage of futurelessness.[54] The 'excruciating...realization that there is no such thing as complete freedom from terrorist danger' can lead some to believe in a 'vision of achieving total sway over human endeavors,' of electing whomever it takes to alleviate the country of the fear that accompanies futurelessness.[55]

The reaction to America's untreated futurelessness, which developed after 9/11 and metastasized after the financial crash of 2008, is an intense nostalgia for previous time periods considered more stable and less fluid than our own. Aided by the archival characteristics of the Internet, which invites users to scroll through most of human history at any time, futurelessness continues to circulate, as fewer people believe they can imagine a future worth living.

Instant Nostalgia

The saturation of Western culture with nostalgia is certainly due in part to Instagram, a social media app that allows users to share photos and videos. Two months after its launch in 2010, the app, now owned by Facebook, Inc., clocked one million registered users.[56] By the end of 2017, that number had reached 800 million.[57] No doubt, the social networking service was immediately popular because of its stock filters that users could apply to images before sharing them. Filters degraded already low-resolution photos, taken with mobile devices, to mimic the photographic fade that can only occur with the passage of time. The experience of scrolling down early Instagram feeds was like flipping through photo albums from lost decades. With Instagram, pre-Recession nostalgia ramped into full form.

Anyone could manufacture their own mediated nostalgia for others to consume.

But the end result was uncanny. Images shot on smartphones wore the patina of Polaroids. New and old, digital and analog, collided, and since the pangs of nostalgia could be assuaged with merely an upload, users presented the illusion of the past to others while still holding on to their digital devices.

Instagram has proven to be adverse for mental health, especially among young people. Like other social media, the app is designed to mete out hits of dopamine to users, a feature that ensures users keep coming back, and to modify behavior on a granular level.[58] It also serves as a platform for bullying, presents unrealistic images of others, and breaches privacy by culling massive amounts of personal information.[59] Hostile invective circulates through the network, as do conspiracy theories.

In an early profile of the app in *The New Yorker*, writer Ian Crouch called Instagram's manufactured sentimentality 'instant nostalgia':

[Instagram] makes everything in our lives, including and especially ourselves, look better. We live in a world of bad lighting, and are forever stuck in mundane locations and posed in unflattering positions. Instagram gives us an ideal self – our edges sharpened by finely tuned manipulations of contrast and color. We look like the subjects of a magazine photo shoot – the nineteen-seventies rock-and-roll stars we always hoped we'd be.[60]

Crouch, writing in 2012, characterized Instagram as a photographic medium that 'rushes and fakes the emotion of old photographs by cutting out the wait for history entirely.'[61] The app certainly facilitates instant nostalgia, and because users can upload as much of their lives as they want to the network, it serves as a reliquary for lived existence. Anything – any event, any moment – can become an instant memory once it's edited and shared on the app.

Much of this obsession with archiving photographic and video material, with spending our time capturing moments as images, was thoroughly critiqued by Susan Sontag, the philosopher and essayist whose landmark work *On Photography* remains a sharp analysis of our relationship with the photographic image. Detailing the social impact of photography, Sontag notes that travel came to be seen as a way to take the best photographs in the most exotic locations.[62] Tourists started traveling to faraway places not to experience another culture, she writes, but to go to work taking photographs. Sontag believed Americans in particular have a difficult time *not working*, even when vacationing.

Writing in the late seventies, Sontag was not the first to connect photography and travel. During his career, Marshall McLuhan – the controversial media theorist from the time of gender essentialism, jazz on wax, hot bomb scares, and Cold War – studied different media technologies, including photography, in order to parse their social, psychological, and cultural effects. Like Sontag, McLuhan understood well how photography impacted tourism and how such an impact turned travel into work. 'People moved by the silliest whims,' he writes in *Understanding Media*, 'now clutter the foreign places, because travel differs very little from going to a movie or turning the pages of a magazine.'[63] Excusing the whiff of superiority (*to be a serious traveler, one must be driven by literary passions!*), it reads like he could be talking about Instagram, some fifty years before its invention.

Instagram makes it much easier to capture our tourism for anyone to see. The end result, however, is a determined motivation to make people look appealing to others, to present versions of ourselves that are happy, cosmopolitan, fit, and popular. But we are not always any of those qualities. I often hear of people taking photographs of their travels to post on Instagram, which earns them more likes and followers. Their feeds are long with edited photographs of their tourist destinations. I remember a friend who took a trip to Norway returning to tell me he wasn't as happy on the trip as he seemed in his pictures. It's hard work getting organic likes and followers. It's often lonely, too.

Instagram has evolved into more than just a reliquary. It still offers stock filters that artificially age images, but it has become less of a museum and more of a highlight reel for many people. Whether or not an uploaded image has been coated in a pre-made filter to make it look like a Polaroid, it still functions as a chosen slice of one's life. And that slice rarely depicts anything mundane or terrible. In order to accrue likes and followers accordingly, the average Instagram user typically posts images that present positive feelings and momentous occasions. Negative reactions, of course, are still common, and Instagram users engage in hostilities as they do on other social media. Memes and weird fare can certainly be found as well, and it cannot be denied that the network connects all kinds of people from all over the world, giving voice to marginalized persons and establishing online communities of sorts. But anything challenging monocultural expectations on Instagram – anything grotesque, mundane, or truly strange that risks falling into oblivion because it's so outré – is often overshadowed by doctored photos of beautiful landscapes and fitness drink advertisements.

An app that principally channels nostalgia, Instagram does not make us feel better. We are often more depressed when we view the world through its lens, even though the nostalgia it offers in spades is, according to some, supposed to be good for our health.

The scientific literature on nostalgia indicates it can 'induce feelings of being loved'; 'counteract loneliness'; 'increase prosocial behavior'; 'enhance self-esteem'; and 'provide a sense of meaning in life.'[64] Social scientists routinely paint nostalgia as a regularly occurring emotion that's quite healthy for individuals. What some scientists define as personal nostalgia, which has been characterized as 'longing for...[a] lived past,' is usually considered a healthy emotional response.[65] Personal nostalgia is like love, pride, and joy insofar as it 'bolsters social bonds... increases positive self-regard...[and] generates positive affect.'[66]

Psychologist Krystine Batcho has studied nostalgia extensively with her Nostalgia Inventory, a 'measure of nostalgia proneness.'[67] In multiple studies Batcho concludes that nostalgia serves a healthy function for individuals.[68] Whereas some scholars consider nostalgia an 'escapist' emotion, Batcho has found no link between it and 'denial, behavior disengagement, self-blame, or substance abuse.'[69] Ultimately, Batcho's research shatters the negative portrayal of nostalgic individuals as 'those who avoid difficulties in their present by seeking solace in fantasy, reminiscence, or attempts to live in the past.'[70] In her view, personal nostalgia is a prosocial, healthy emotion and not an obsessive longing to live in a former time.

Pre-Recession nostalgia is quite different from the personal nostalgia Batcho and other social psychologists study, simply because it is both personal and historical. The nostalgic sufferers of today often think of their own personal nostalgia as imbricated with a larger yearning for another time period, perhaps one they never lived through. And even as personal nostalgia sometimes enables prosocial behavior, its commodified form mostly does the opposite. The nostalgia of Instagram – this presentist 'instant nostalgia' designed to sell products and ideas – is regressive and ahistorical. Those who yearn desperately to escape the present see different utopias in the past: perhaps a time before smartphones or student debt, maybe a whitewashed society before the civil rights movement.

Nostalgia doesn't have to be reactionary. We can learn a great deal from the past to improve the present. To chart a path towards greater change, we can look to times when there were more public services, before billionaires wrote the political script, before tech companies started manipulating behavior, and so forth. We must take actions to do this, not hide our heads in history's quicksand. And those actions must better the lives of all people, not deprive some of their humanity while privileging others. But the pre-Recession strain of nostalgia is not being used in this way. Rather, it teaches the public that things were always better in the past.

Chapter Four

In the Mirror Maze: Regression, Recursion, and Canonization

The way it was, was right. The way it is, is wrong.
- *Ruth Bader Ginsberg*[1]

The machine rotates on the same spot.
- *Max Horkheimer and Theodor W. Adorno*[2]

Judging by the glut of children's products aimed at adults, it seems much of the public today wants to relive the breezy wonder years of youth. The more regressive branches of the nostalgia industry sell products designed to soothe anxiety and ease the daily pressures of adulthood. Citing research on the calming effects of nostalgia, the products promise relaxation and peace of mind in a distracted age. They also encourage the idea that recovering the past will always improve the present.

When Johanna Basford's coloring book *Secret Garden: An Inky Treasure Hunt and Coloring Book* was released in 2013, it quickly assimilated into an already growing industry of children's activities marketed to adults, including adult preschools and digital detox camps. This 'Peter Pan market,' or what is sometimes called the kidult market, rose from within the nostalgia industry and has become quite successful at peddling regressive wares.[3]

Having sold eleven million copies, *Secret Garden* struck a chord with a stressed-out world. Economic precarity and digital saturation combined with the routine pressures of adulthood aided the popularity of hers and other adult coloring books. Desperate for relief from a complex, unstable world, many have taken to activities typically reserved for children. They believe childhood was a period of less stress and greater freedom.

News platforms and online aggregators alike promoted the positive effects of the kidult market, papering over many of its criticisms. *HuffPost* provided two articles in particular advancing claims that coloring books aid in self-expression and stress relief.[4] In one, journalist Priscilla Frank interviews Johanna Basford, who admits that nostalgia is the central reason behind the popularity of her book. 'Chances are last time you spent an hour or so coloring…you didn't have a mortgage,' Basford says, 'and you weren't worried about a nagging boss or the financial crisis!' By the article's end, Frank recommends readers to '[c]onsider trading in your yoga mat for a set of markers.' In another article, Frank compiles a list of the ten best coloring books for adults. She begins by writing:

> We're fairly certain elementary school had all the answers. Nap time, snack time, recess, second nap time — how could any third-grader not be completely zenned out? Another major factor in our eternal envy of the single-digit age group: the coloring book.

Like others working in the nostalgia industry, Frank makes a fundamental error about childhood and the past in general. Children aren't 'completely zenned out' – or, at least, they probably shouldn't be. Children vacillate wildly between mania and rest, and childhood is often as distressing and boring as it is carefree. Adult coloring books encourage us to have little appetite for anything new and to tend towards compliance – what Sherry Turkle calls 'the sweet spot of simulation: the exhilaration of creativity without its pressures, the excitement of exploration without its risks.'[5] There is little risk in a society where adults mimic the so-called 'zenned out' nature of children.

Other kidult products and services soon followed. Adult preschools, like Michelle Joni's Brooklyn-based 'Preschool Mastermind', promised to teach 'grownups…the basics, and experience the magic of life as it was originally intended.'[6] For a fee, adults could finger paint, nap, play games, and attend field trips.[7] In 2016, the augmented reality video game *Pokémon Go* exploded into the mainstream and quickly amassed millions of downloads.[8] Released some twenty years after the first generation of *Pokémon* games, *Go* capitalized on the memories of late twentieth century 'Pokémania,' this time replacing Game Boys with smartphones. And in 2016, Netflix released its kidult series, *Stranger Things*.

Childhood's In

Created by Matt and Ross Duffer, *Stranger Things* is a streaming series about a group of young teens who battle against the forces of evil in the early 1980s.[9] Drawing heavily from *The Goonies* and *E.T.*, the series has had massive success, spawning everything from memes, comic books, and Lego sets. Even singer-songwriter Ingrid Michaelson paid extended homage to the series by writing an entire album, *Stranger Songs*, inspired by its characters and storylines.[10]

Stranger Things is set in the 1980s in Hawkins, Indiana, a fictional small town that exists at the center of supernatural events. The main cast of characters must confront the monsters and mayhem that threaten the town, all while navigating the rocky waters of adolescence. There are government conspiracies, an alternate dimension, medical experimentations, and psychokinetic children.

But the series is about more than young love, high school, and cross-dimensional turmoil. *Stranger Things* is also a showcase for the bygone popular culture of the 1980s. Old-fashioned technology and iconography fill every inch of the screen. The one-dimensional characters pale in comparison, playing secondary parts to the images of the past, which are really the stars of the series.

A prodigious amount of research went into recreating the

1980s during the series' pre-production. With the guidance of prop master Lynda Reiss, the Duffer brothers wrote the premise of the series and much of the first season around the 80s props they wanted to feature. In a 2016 *Wired* article all about the 'period-correct gear' featured in the series, Reiss admits she didn't want *Stranger Things* to be 'nostalgia-tinged.' 'I want it to *be* the '80s,' she says. 'I don't want it to be what everyone just *thinks* is the '80s.'[11]

For Reiss and the Duffer brothers, this meant refraining from picking only the most obvious pop culture markers of that period. So they went further and mined the most obscure, mundane, and everyday corners of the American 1980s to create a show not merely tinged with nostalgia but distended with it.

Two events kick off the first season: the disappearance of Will Byers, a twelve-year-old citizen of Hawkins, and the appearance of Eleven, a nearly mute young girl who wields psychokinetic power and harbors dark secrets. In a scene from the season one episode 'Holly, Jolly,' Eleven explores the house of Mike Wheeler, a young boy and friend of Will's. Eleven reclines in a cushioned La-Z-Boy, hears for the first time the dial tone on a landline, and surfs the channels on a CRT TV. She peers in wonder at them while pausing to inspect her surroundings. We the viewers encounter these pieces of furniture and technologies along with her, gawking like visitors in a museum of the 1980s.

In a lengthy scene in 'The Weirdo On Maple,' Joyce, the mother of the missing Will Byers, installs a new telephone after an unknown force short circuits her old one. We witness her unbox the phone, untangle the cord, and plug the cord into the jack. She then tries to walk into her living room holding the phone, but the cord catches her. Confused, she looks around at what has stopped her and tries to yank the phone away from whatever is holding it. Realizing the phone cord is too short, she scoots a chair closer to the jack and sits. It's as if she were a twenty-first century smartphone user flung back in time, baffled by a phone with a cord connected to the wall. In a show less reliant on nostalgia, the scene would have likely been edited, or perhaps left on the cutting room floor. But it functions primarily as a curiosity exhibit for twenty-first century viewers. Even in this derivative version of the past, she dreams of the freedom of movement that mobile phones provide.

When the Duffer brothers realized the Bell Telephone Company monopoly wasn't dismantled until after 1981, the first season's setting was changed from 1981 to 1983 to account for the phone scene.[12] It's unclear whether this kind of presentism is likely inevitable whenever content creators – and everyone else – attempt to recreate the past. One might assume that to remember the past is an act of presentism itself. But presentism isn't merely misremembering; it's also an act of exclusion. What gets remembered and re-created silences what gets lost. The narratives written from the dregs of dredged-up memories become the monolithic authority on history. When those memories neglect the actual lived experiences of others, the result could be erasure.

And when we assume the figures of the past simply dreamt of the technologies to come and perhaps even suffered from the limiting technologies of their day because they *knew* what was coming, we risk thinking of them as naïve, innocent, even lesser. It's quite easy for some to have pity on the denizens of history. The privileged few, from the comfort of an air-conditioned tower, can afford to gaze sadly at the richest kings of antiquity, who shared little of the amenities that the modern world has afforded. But such a perspective ignores the role of power in history. Air conditioning and mobile communication may seem heaven-sent compared to sweltering heat and corded phones but only if you have such comforts today. Kings were kings in ancient days, even without sanitary water and plumbing, and many today still suffer, even though smartphones exist. Believing the present is the best of all possible worlds is a Panglossian mistake, one as destructive as thinking the past was a utopia. History is far more complicated than that. The people of the past were better off in some ways, and in others, not so much. But when we simply plunder bygone references from history's naïve denizens, we risk reducing the past to a well from which we draw resources to enliven otherwise parched myths.

Watching *Stranger Things*, many viewers wish to live in its simplified world. As much as they might love the convenience of smartphones, fans of the series yearn for landlines and early cable TV, as well as the small town in which the show takes place. It's a town populated by working-class people who live with a sense of community, where kids can socialize without apps, where citizens know the names of local police officers, and where the good guys always win. Without social media, climate crises, and mass shootings, the characters of the show are freer than the viewers watching.

Lieux de Mémoire

Gwinnett Place Mall in Duluth, Georgia, was once a hugely popular destination for area shoppers in the 1980s. At its height, it was the second highest-selling mall in Georgia and attracted patrons from across the state. Decades later, the mall is now largely defunct due to the rise of North Georgia megamalls and online commerce. Much of it is in disrepair, and almost all of the anchor stores have fled. A handful of stores still remain, but the mall is eerily empty on most days. In December 2017, a young woman was found dead in the food court, and investigators believed she had been there for nearly two weeks.[13]

For the third season of *Stranger Things*, the creators chose Gwinnett Place Mall to serve as one of the central locations. The massive renovation project transformed the failing mall into Starcourt Mall, the latest shopping development in the fictional Hawkins. In a fake advertisement for the third season, Starcourt is touted as a family-friendly spot to shop, eat, and hang out.[14] The teaser video resembles the mall ads of the 1980s, and Starcourt itself is a near-perfect replica of early malls, where shoppers could purchase clothes at The Gap, books at Waldenbooks, and physical music at Sam Goody. For the length of the shoot, Gwinnett Place Mall was briefly transported back to its heyday.

Even before it was given a nostalgic facelift, Gwinnett Place Mall was a lieu de mémoire, a site of memory people could visit to travel back in time. Like monuments and museums, the mall was touched with the residue of the past, making it a kind of stigmatized property stained with history. Its emptiness and disrepair merely intensified its dull glow, imbuing it with an anachronistic aura. Both alive and dead, situated in the present and the past simultaneously, sites of memory like Gwinnett Place exist in a liminal space. Walking through the mall, you know it's a relic from a bygone time, but enough of it is still intact to nearly zap you back to the 1980s.

Its transformation into Starcourt Mall resurrected this site of memory. The possibility that the mall *could* return to the 1980s came briefly true, that if enough of the past can be recreated, right down to the most minute, inconsequential details, then perhaps memory sites can come to life.

A similar resurrection occurred during the filming of Netflix's 90s high school series *Everything Sucks!* In a scene from the eighth episode of the first season, teens Kate and Luke visit their local Blockbuster. Instead of building a fake Blockbuster, the film crew shot the scene in one of the last standing Blockbusters in the U.S., located in Sandy, Oregon.[15] To fit the setting, the crew gutted the store of DVDs and Blu-Rays and stocked the shelves with VHS tapes, reviving the site of memory in much the same way that Gwinnett Place Mall was. But the resurrection ultimately failed; like other Blockbusters that were wiped out in the wake of Netflix, the Sandy location closed immediately following shooting.[16]

Sites of memory can be crucial for coping with the passage of time as they allow communities to reimagine or recapture what has been lost. But they also run the risk of objectifying the past as a kind of theme park we can visit, a place where history *really is* and where we can refuel on true meaning during current crises of authenticity. They are also sites of oblivion subject to selective memory, where remembering dovetails with social forgetting.[17] When they are erected, they risk obliterating certain ideas and peoples.

Nostalgic Nightmares

Sometimes when we revive the dead, it comes back more horrific and disfigured than we remember. This is portrayed in David Robert Mitchell's 2014 take on the stalker film, *It Follows*; a frightening story about a nightmare without end and a blank terror that cannot be thwarted. Far more nuanced than *Stranger Things*, it is a film about our fear of the present.

At first glance, the plot of *It Follows* scans as a textbook example of the stalker genre. After having sex with her new boyfriend, Jay, the film's protagonist, is followed by a ruthless and extremely violent entity. The entity, which assumes numerous human forms, can only be stopped if Jay has sex with someone else, passing on whatever it is that draws the entity closer. If it catches her, it will kill her and resume following the last person she had sex with.

It Follows can be considered a fable about the dangers of sexually transmitted infections because sex is the means by which the mark of death is passed from character to character. But there are other characteristics of the film that invite different interpretations.

First, the film is set in no distinct time period. There are stylistic markers from the American 1950s to the 1990s, but the setting is completely timeless. The soundtrack, scored by electronic artist Disasterpeace, borrows generously from the iconic synth scores of John Carpenter, especially his compositions for the *Halloween* franchise. And the film poster design recalls the horror posters from the 70s and 80s. Yet there is no explicit reference to what decade the film is set. Second, parents or guardian figures are noticeably absent from the film. The few parents who are featured are inept, skeptical, or silent. Most of the adult figures are actually manifestations of the murderous entity. And finally, the technologies in the film are all dated: twentieth century automobiles, television sets, phones and so forth. With the exception of a shell compact that doubles as an e-reader – a piece of technology that has never existed – much of the tech in the film pre-dates the twentieth century.

A film about adolescent struggle in a dreamworld knocked out of time, *It Follows* is also a metaphor for the youth experience in the West. On its surface, it appears to capitalize on the 80s craze of the 2010s, but it is more progressive than the conservative dream presented in *Stranger Things*. It depicts a world where authority is useless against the ambulatory force of oppression targeting young people. It reflects the anxiety of becoming adults in the twenty-first century, when the cultural detritus left behind from previous decades papers over the disasters of today: the crushing tonnage of student debt, economic instability, total war, crumbling social services, and environmental collapse. The entity in *It Follows* is the culmination of those problems handed to younger generations, and for many both in and outside the film, it seems there is no way to defeat the monster.

However, the nightmare world of *It Follows* is actually far simpler than the one young people have inherited. Its world is one without smartphones and ubiquitous digital technology. There is no Facebook or Google or Twitter. There is no Apple or Amazon. There is no online forum for victims of the entity to help each other figure out how to stop it. Though bound tightly with each other, the characters in the film are detached from the wider world. It is either a world living after the collapse of Big

Tech, presented as a crumbling bolus of poverty, or a dream of a society untethered.

In an interview with David Robert Mitchell, he notes that he didn't include smartphones in the film because doing so would have 'date[d] it.' Tellingly, he admits a smartphone is 'too real for the movie.'[18] The exclusion of contemporary digital technology lends the film its timeless aesthetic, yet it also suggests we are living in a time when technology is so powerful and pervasive that nothing, not even our imagined horrors, can outsmart it. This is, of course, a dangerous myth, for if we believe Big Tech is an unstoppable colossus, then there is no use in resisting its authority. It seems the entity, whether it symbolizes STIs, the inherited socio-political problems of our time, or even the unrelenting logic of capitalism, could only exist in a time before smartphones.

To manage this tension between plot and technology, some content creators covertly inject mobile communication into their narratives. The plot of the second season of *Stranger Things*, for example, is determined in part by the characters' ability to communicate with one another via walkie-talkies and radios. Because smartphones didn't exist in the 1980s, the Duffer brothers cobbled them together using the extant technology of the time period to assure the plot ran smoothly. The result is a world where micro-coordination is possible, where characters have some control, but not the total control smartphones allow.[19] In *It Follows*, the characters have no such control.

Perhaps those of us watching *It Follows* yearn for a time when our monsters were harder to defeat, when it seemed the protagonist was no match against something evil and an entire movie was dedicated to stopping the monster. Instead, we have been given a means of controlling nearly every aspect of life, of knowing everything all the time, and as a result, entire narratives are rendered incoherent. How can we imagine horror in a time when everything can be known? Or when daily life is spent outrunning inherited horrors?

Of course, there are plenty of innovative horror films being made in the digital era, but it cannot be denied that narrative storytelling – no matter the genre – has changed along with our lives since the invention of mobile computing devices. It seems the more powerful our handheld devices are, and the more tethered we are to them, the more we dream up worlds without them.

Recursion

Once it became clear that pre-Recession nostalgia could sell culture, content creators took note. The more we bought and streamed nostalgic media, the more our digital platforms offered nostalgia on tap. Thanks to the archival structure of the Internet and the instability of a precarious world, we easily fed nostalgia into the streaming algorithms that were popping up at the turn of the 2010s. The result was, and continues to be, a streaming model that feeds on and regurgitates nostalgia.

In an attention economy, nostalgia sells.[20] When attention is diffused across multiple media, it becomes a valuable target for advertisers and marketers. The attention of American individuals, in particular, is constantly being wrenched from mobile device to computer screen, Facebook to Snapchat, Amazon to Spotify. This says nothing of the ways our attention is diverted offline as we stroll through brick and mortars, drive automobiles down billboard-lined roadways, and engage with more traditional media like television and radio. It can be difficult to parse what is advertising and what is not. In this kind of control society, a universal emotion like nostalgia sells very well.

Nostalgic branding has proven to be an effective means of selling products. A 2013 study conducted by Altaf Merchant and Gregory Rose at the University of Washington found that content creators from filmmakers to corporate advertisers often employ nostalgia to forge bonds between consumers and their products.[21] They noted that during periods of crisis, nostalgic branding can 'reinforc[e] perceptions of stability, creat[e] positive emotions, and communicat[e] the consistency of the brand's promise over time.'[22] By appealing in these ways, the brand has the consumer's attention, and since attention is being pulled in several directions in a technologized society, content creators must work tirelessly to secure a single individual's attention. Tech companies do the exact same.

In 2006, Netflix announced the Netflix Prize, a contest worth one million dollars that 'challenged the data mining, machine learning and computer science communities' to come up with a more accurate recommendation system for the company.[23] Netflix publicly released the movie ratings of millions of subscribers and scrubbed them of any personally identifying information.[24] In 2009, a team of scientists from AT&T called BellKor's Pragmatic Chaos won the prize, but the contest was not without its problems. In February 2008, two researchers from The University of Texas at Austin published a paper revealing that by cross-correlating the anonymous subscriber ratings released by Netflix with other information sources, such as the Internet Movie Database (IMDb), an individual could easily determine the identities of those subscribers.[25] This revelation led one woman to sue Netflix for invasion of privacy, forcing it to abort its plan to offer a second prize.[26]

Yet Netflix has not backed down in its relentless quest to better recommend and even predict user preferences. Like Amazon and Spotify, Netflix operates by inviting users to input preferences and providing content that matches them. Handing over private information eventually becomes normalized as we pursue more content.

This process of providing and predicting based on personal preferences is called collaborative filtering, which media theorist Alexander Galloway describes as such:

A user answers a series of questions about his or her likes and dislikes, thus setting up a personal 'profile.' The profile is entered into a pool of other profiles supplied by other users. Then, statistical algorithms suggest other likes and dislikes of the user, based on the similarity of his or her profile to other users' profiles in the pool.[27]

What the researchers at The University of Texas at Austin revealed in their paper was essentially Netflix's strategy: cross-correlate multiple sources of information to target users on a granular scale. Over time, Netflix and other streaming giants perfected this technique, thus creating a kind of cultural feedback loop. Once the preferential information is entered and compared with others in the system, Galloway writes,

> no improvement in the overall data pool is possible. Thus, collaborative filtering ensures structural homogeneity rather than heterogeneity. While any given user may experience a broadening of his or her personal tastes, the pool at large becomes less and less internally diverse.[28]

The result of collaborative filtering is a hall of mirrors, each mirror reflecting the tastes and desires of one's own, which are themselves reflections of someone else's. 'Personal identity is formed only on certain hegemonic patterns,' Galloway writes. 'In this massive algorithmic collaboration, the user is always suggested to be like someone else, who, in order for the system to work, is already like the user to begin with!'[29]

Out of this hall of mirrors come cultural artifacts, like movies and music, that are also forged on fixed grids, and as algorithms continue to deliver more of the same, users start to desire the culture that continues to affirm their tastes. Musicians, filmmakers, public relations agents, and other industry professionals know this is how the game is currently played, so they capitalize on the feedback loops generated from collaborative filtering. And in search of subscribers, companies like Netflix and Amazon give sizable budgets to industry professionals to create and star in streaming series that remove the middleman from the equation. Serving as both studio and aggregator, Netflix can engineer content based on user preferences and fold its original programming into the slew of content available from third-party producers. It is a tactic to give viewers what Netflix thinks they want.

Not surprisingly, if nostalgia already circulates widely in culture and streaming giants offer content based on user preference, then nostalgic content will continue to circulate. Even if nostalgia were absent from Western culture, the algorithms designed to suggest new content would inevitably recommend music and movies that resemble prior preferences. In other words, nostalgia is baked into the very recommender systems that structure streaming providers.

Changes in taste and the thirst for novelty are written out of the codes underpinning such content platforms. The more our information is gathered by data collection companies like Acxiom, Datalogix, and CoreLogic, the more our desires are affirmed. And when predictive analytics determine taste, the result is a culture of recursion.

Ours in the twenty-first century is as much a recombinant culture as a recursive one. In a recombinant culture, media objects glom together to create a kind of pastiche. For example, pop songs that sample other tunes are fixtures in a recombinant culture. So are crossovers, sequels, and spin-offs. But once advertisers and media corporations started employing more finely tuned recommender systems, creative content began cycling over with astonishing speed, thus leading to a culture of recursion. Thanks to streaming, the near-ubiquity of Wi-Fi, and the ability for users to process and generate content more easily, cultural products are often recycled quickly in the digital age. And each time a cultural object is spat out the end of an algorithm, it's more derivative than it was before. Even if we wanted to escape the cul-de-sac of nostalgia, doing so would be at odds with the economic model of algorithm-based streaming.

The reality is, you can't really predict a person's behavior, even with sophisticated analytics. So instead, with the help of collaborative filtering, culture repeats and settles into a homogenous swirl, and the results merely appear as prediction. Stuck in feedback loops, much of the art we consume runs over the very same ground in cycles, making it easier for Big Data to 'predict' what we want next.

The paternalistic gesture of being offered things homologous to our supposed desires seems ideal, but it is a siren song. Instead of routing consumers towards new ideas, recommender systems invite them into mirror mazes, each turn revealing reflections that recede to infinity. Like the dizzying and bottomless Mirror Maze in Ray Bradbury's *Something Wicked This Way Comes*, modern recommender systems show us derivatives of ourselves, each mirrored image more alien than the last. An attraction at a dark carnival that lures customers in by promising immortality, the Mirror Maze in Bradbury's novel is appealing at first sight, perhaps even fun. But upon entering, the carnival-goer is doomed. In Bradbury's lavish prose, he describes a character's reaction to the maze:

> Ahead flowed sluices of silver light, deep slabs of shadow, polished, wiped, rinsed with images of themselves and others whose souls, passing, scoured the glass with their agony, curried the cold ice with their narcissism or sweated the angles and flats with their fear.[30]

A documented bigot who believed minorities are as threatening as totalitarian regimes, Bradbury might not recognize the mirror mazes of the present century if he were alive to see them.[31] Yet he knew they would ultimately destroy a person, for who could bear the sight of their own derivatives – copies that dumbed down the deeper you peered into the sea of reflections? In the Mirror Maze, the old look older; the young, younger. New images never materialize. There are only endless throngs of simulations.

Canonization

As Western culture both regresses and recurs, content creators search for new takes on old ideas. After exhausting more obvious cultural references from the past, they scan uncharted historical areas to colonize. A cultural object like a film is dug up, and elements are added, refined, and updated until the film becomes its own universe. The upshot is not only a deeply nostalgic society but also one in which stories function as worlds in which we are invited to live. With prequels, sequels, reboots, and reimaginings, narrative media promises total immersion by expanding widely on original stories. Prequels enrich plot arcs by providing backstories. Reboots and sequels introduce new characters while shedding light on the psychologies of older ones. Reacting to such universe expanding, ardent fans debate the authenticity of re-creations – do they stay true to the original vision while also keeping it fresh?

Too often, these expansions jump the shark, leaving creators and fans with a predicament. How can filmmakers continue building the fictional universes of our favorite franchises without sullying the originals and, in doing so, losing money? And what do we do with those entries deemed critical failures or box-office flops?

For many, the answer is to build canons in order to construct coherent narrative universes while paying homage to the originals that started it all. Canons tell old and new fans alike which creative works matter, and which don't. Knowing a canon primes viewers for the latest reboot. That which is non-canonical is considered supplementary and, ultimately, nonessential. The canon building of the Marvel and DC series, the *Transformers* franchise, *Harry Potter*, *Alien*, and, of course, *Star Wars* (among so many others) gives viewers simplified worlds in which to live. And by creating canons, we as fans are invited to control the past.

One of the franchises to reboot in 2018 was *Halloween*, which began in the late-1970s as a meager low-budget slasher film directed by genre *auteur* John Carpenter. Going on four decades, the *Halloween* franchise has expanded from a lean horror film with a simple premise to include telepathic children (*Halloween 5: The Revenge of Michael Myers*) and ancient Druid curses (*Halloween: The Curse of Michael Myers*).

After killing off Michael Myers in *Halloween II* (1981), Carpenter hoped to continue the series by featuring a new threat for each film, a move that would have turned *Halloween* into an 'annual horror anthology,' but his efforts proved fruitless when he underestimated the avid fandom surrounding Michael Myers.[32] After an outpouring of negative backlash against *Halloween III: Season of the Witch* (1982), which did not feature Michael Myers as the villain, Carpenter and co-producer Debra Hill signed away their rights, and a production cleanup crew led by Moustapha Akkad revived the Myers character in *Halloween 4: The Return of Michael Myers* (1988).

In an attempt to scrub the series of its more harebrained plot twists, *Halloween H20: 20 Years Later* (1998) ignored every sequel after *Halloween II*, but it and its own sequel, *Halloween: Resurrection* (2002), were also critical failures. Later, Rob Zombie revived the franchise with two regrettable nonstarters, *Halloween* (2007) and *Halloween II* (2009), that relied more on guts and less on the nuance and suspense that made the original *Halloween* memorable.

At this point, the *Halloween* series has become a nearly untenable exercise in both expansion and preservation. Its revivers attempt to manage the tension between making a buck and trying not to totally destroy the magic of the original film – in other words, between fan service and integrity.

The eleventh film in the franchise, *Halloween* (2018), was another Hollywood maneuver to cash in on the original's legacy without acknowledging the various sequels and their offbeat narrative developments. The film ignores the entire franchise, save the original that started it all. Without the messy sequels and derivative reboots standing in the way, filmmakers Danny Gordon Green, Danny McBride, and Malek Akkad were lent the freedom to start anew while also giving credit to the original film.

Much of what the *Halloween* filmmakers, fans, and critics talked about regarding these creative decisions was the idea of building canons. By ignoring every *Halloween* film but the first, the filmmakers and fans had to adjust to a new *Halloween* canon that only includes the 1978 original, the 2018 sequel, and any other installments to come. The conversation over what does and doesn't comprise the canon mirrors other debates about *Star Wars*, a franchise with a sprawling universe that invites extensive discussions of canonicity. But still, the question must be begged: why do we even canonize pop culture?

Western culture is concerned with building canons because doing so enables fans and content creators alike to preserve and, by extension, control the past. In an effort to keep the franchise bloodline pure, borders are drawn separating what is and isn't canonical.

But like all canons, erecting them leaves out the things that don't fit: the uncommon things, the ideas and voices locked out of the conversation by fiat. Canonization erases those parts of the past deemed unworthy, dangerous, or embarrassing. Western literary canons privilege certain voices – typically white, heteronormative, male perspectives – and there is often debate over who gets included. Determining which ideas are 'great' requires one to exercise power over the innumerable ideas considered less so. By deleting much of our past, canon building simplifies history to a digestible story of progress starring those figures eligible for inclusion. The remaining figures are hidden from history because they do not comport with the criteria of the canon. Ultimately, when we build canons, we commit violence by marginalizing the perspectives that some believe threaten the flow of progress.

In our contemporary media environment, the outlier films and narratives are not so much erased as often reduced to Easter eggs, which we, as good fans, are supposed to hunt and collect while watching the canon. Astute viewers identify the hidden allusions and tally them. Such a skill separates the franchise fans from the casuals. Although the 2018 *Halloween* entry ignores much of the *Halloween* universe, it includes several Easter eggs that 'salute' the previous 'mythologies,' a decision that essentially reduces the majority of the *Halloween* franchise to footnotes.[33]

The popularity of Easter eggs in media has increased tremendously in the twenty-first century. A cursory search yields countless videos and listicles cataloguing hidden references throughout mainstream Hollywood films, perhaps none more so than *Ready Player One* (2018). Although not a franchise (yet), *Ready Player One* is assembled largely out of Easter eggs and boasts, at the center of its narrative, one very important Easter egg.[34] The film built its own canon by cherry picking the parts of pop culture deemed worthy of inclusion in our cultural imaginary and presents a virtual version of previous decades (mainly, the American 1980s) that ignores anything countercultural. It is a film, as critic Richard Brody notes, depicting a past 'without hip-hop, without punk, without Patti Smith.'[35] Director Steven Spielberg's purposeful exclusion of anything potentially incendiary encourages us viewers in our surveillance society to be good citizens, refrain from stirring the pot, and stick to hunting eggs.

Spielberg, along with his fellow fantasist George Lucas,

has written much of the cultural script for the past forty years, and *Ready Player One* is a revision of cultural history, written Spielberg's way, without the messier parts. It's a direct reflection of our control society, in which the ubiquity of digital technology provides the illusion of freedom. But we're still playing someone else's game, and as long as we're hunting eggs, we won't question the state of things.

By reducing to Easter eggs those *Halloween* films with lower Rotten Tomatoes scores, Green, McBride, and their team (including John Carpenter, who approved of de-canonizing all of the sequels) provide us with a clean franchise and an orderly film-world.[36] Watching future *Halloween* installments, we can judge whether they, too, will be worthy of canonization. And like diligent analysts, we can hunt for clues that allude to those messier parts without being bothered by their silly subplots, over-the-top left turns, or dated features.

But what we need aren't canons that correct the wrongs of the past, because any canonical gesture automatically engenders exclusion. Instead, we should abolish canons. Without them, we can promote a more complex cultural history. It's one of the better options we have left during this long period of nostalgia-baiting.

'An Authoritarian Alternative'

The scientific literature claims that nostalgia often promotes prosocial behavior, but it also can be (and has been) used to promote ideas about how we might live. Sometimes those ideas skew toward the reactionary, perhaps even the authoritarian. Pre-Recession nostalgia, with its eye cocked to the 1980s and early 1990s, often carries with it a nativist, nationalist element that can be induced to mobilize some groups of people against others. Reacting to the instability of American life during a prolonged global war and after a crippling financial recession, some publics see nostalgia as the escape route. They are being told that the enemies who caused such instability are pathetic establishment politicians, immigrants, and leftists. Certainly, not every American today is nostalgic, and not every nostalgic person in the country today is a Trumpist. But it is the case that nostalgia, which peaks during unstable cycles in history, has proven to be a politically powerful tool to unite some and divide others.

In 2001, rhetorical scholars Robert Hariman and John Lucaites addressed this very issue in their essay on John Filo's iconic photograph of the Kent State shootings on May 4, 1970. They determined that some 'emotions are...good models for political education' but saw in modern politics a general suppression of emotions, which, they claimed in the essay, are healthy forms of dissent and should be taken seriously.[37] However, they warned that emotions, and emotion-inducing visual imagery in particular, could 'reinforce an authoritarian alternative' in politics.[38] Although emotion can be employed as a force of dissent, it can also give rise to strongmen who might promise to stabilize an emotionally distraught public using a variety of tactics.[39] These leaders might protect the country in ways that are less 'consistent with democratic ideals.'[40] They might use violence. Whatever their methods, and however potentially heinous those may be, these kinds of leaders could act unilaterally in the name of protecting a 'fragile, vulnerable' public.[41]

Donald Trump proved to be exactly the kind of leader of which Hariman and Lucaites warned. He declared America broken, impotent, and dead. And he charged himself fully capable of rejuvenating the country with a variety of methods. One such method was to induce nostalgia for an Edenic period in American history. As he imprisoned migrants and endorsed white supremacists in the name of making America great again, he further demonstrated that nostalgia can be weaponized.

Nostalgia is one very important emotion that travels roundly in the vast orbits of digital media in the twenty-first century. It is felt by individuals but appears with great frequency in the channels of media consumed by most Americans on a daily basis. In its mediated pre-Recession manifestation, it tends to misrepresent the past in order to encourage the public to view the present as unfixable or, worse, in need of an authoritarian figure to police citizens. Ultimately, this popular manifestation of nostalgia encourages widespread amnesia during a time when historical awareness is badly needed. Looking to the past can certainly aid in constructing a better future, but when that past is depicted as sanitized, the result may be troubling. The turn towards nostalgia might make the public feel good. It might push individuals to seek community with others. But it is currently being used to spread ideas that might not align with the goals of democracy. The version of history many yearn for is one that existed in pop culture and on screens in the years before the present century. It is, however, not the history that existed outside its media representations, and as a result, it limns an amnesiac's rendering of the past, grabbing attention and selling ideas about what's worth remembering and repeating again.

Chapter Five

Mending Broken Windows: Virtual Reality and the Dream of Control

Repetition is another way of silencing.
- *Lorgia García-Peña*[1]

This could be heaven.
- *Public Image Ltd., 'No Birds'*[2]

Nostalgia is an emotion of control. What we participate in when we reboot the past is a desperate mission to control history – to revise, relive, or possibly reclaim it. To achieve this, the individual suffering from nostalgia may do whatever it takes to repossess what has been lost. But as nostalgic people yearn to control time, so too are they controlled.

Social media is inherently nostalgic because it allows users to control their history. We do this by posting and sharing what's happening right now, which gives the impression that we are living in an eternal present. But much of what we post online accumulates in some form. We leave digital remnants behind for us or someone else to see. And in fact, social media is almost always about sharing what has already happened: capturing events, editing them to suit our desires, and presenting them in ideal forms. An act of time travel, scrolling through a user's feed lets us see a given person's immaculate surrogate. The feed is an historical document, but it lies. It is a manicured history.

In part, this is why it's shocking when someone digs up an older, problematic post and publicly circulates it. The damning post in question has been overlooked. With so much power to control our own past, how could someone ignore something so heinous? The fact is, even with the power to control time, to curate and delete what does not belong, the ugliest parts of being human always slip through.

When I talk with others about the problems of social media, I often hear a similar refrain: they like social media because it connects them with other people. Without it, they couldn't keep up with friends and family. But keeping up with loved ones and connecting with them are entirely different. It only seems like social media allows connection when in fact it enables voyeurism and fuels the desire to control.

In 1990, French philosopher Gilles Deleuze wrote 'Postscript on the Societies of Control.' The short missive outlines what he saw as the shift from disciplinary societies, as theorized by Michel Foucault, to control societies. In a disciplinary society, people move from one enclosed space to another – from the home to the school to the factory.[3] Each space is separate and distinct, and an individual leaves one before entering another. Writing one year after the fall of the Berlin Wall, Deleuze believed all disciplinary institutions were reaching their twilight and that further reform to fix them would do no good. The remaining disciplinary organizations would be wiped away with the coming institutions of control.

Under control, things are fluid and constantly shifting. The factory, which is fixed and immovable, is no match for the corporation, which, Deleuze writes, is 'a spirit, a gas.'[4] Where the factory fought to make bodies docile, to quell insurrection and competition, the corporation encourages competition, with winners receiving higher salaries. When Deleuze wrote the essay, this model was already spreading throughout Western societies and infecting the former disciplinary institutions. The school, the hospital, and the prison would all adopt the corporate model.[5]

What is left in the control society is the illusion of freedom: the open highway, the gig economy, affordable goods Made in China™. But there is, in reality, little more than marketing in a control society. Nothing really gets produced. The production occurs instead in peripheral countries and the Global South, far away from the eyes of the consumer.

Deleuze's 'Postscript' reads like a comedown from his previous work. His mid-career writing can be thought of as philosophy without borders – a manic but ethical nomadism ignoring divisions between high and low cultures. He, along with his collaborator Felix Guattari, hated the determinism of psychoanalysis and advocated for the dissolution of what they called 'striated space.'[6] They wrote about these and numerous other ideas in *Capitalism and Schizophrenia*, a two-volume opus in which he and Guattari coined many of their most noteworthy concepts. In the first volume, *Anti-Oedipus*, the two thinkers called for the emancipation from capitalism, which induces neuroses and oppresses populations. Nearly a decade after its release, they collaborated again on the second volume, *A Thousand Plateaus*. In it, they argued for a rootless existence, one in which speed, movement, and formlessness are encouraged. There should be no foundations, Deleuze and Guattari argued. To live freely, they thought, we must never settle, always be on the move, and live as if we're in the middle of things.[7]

What Deleuze and Guattari didn't imagine during their time writing *Capitalism and Schizophrenia* is the exhaustion that could come from living a rootless life. Not everyone wants to jump from point to point without any center. They failed to foresee the problems that would develop from never planting roots.

Many of Deleuze and Guattari's concepts were taken up in unusual or antagonistic ways through the end of the twentieth century and into the present.[8] The very structure of post-Fordism resembles their theories of rootlessness (*Work from home! Work on the go!*). In *Anti-Oedipus*, which was partly a reaction to the failure of the May 1968 strikes in France, Deleuze and Guattari advocated for the anarchic acceleration of capitalism.[9] By the early 1970s, they believed capitalism had become so naturalized that many thought of it as the only economic option. Deleuze and Guattari agreed; there would come no alternative.[10] But by the early 1980s, they dialed back their accelerationist tendencies. Too much capitalist acceleration might be horrible, they thought. The consequences could be a more sinister, more exploitative, and more thoroughly entrenched form of capital neither could predict. By the time he wrote the 'Postscript' in the early 1990s, Deleuze had seen the error of his thoughts: a rootless existence is exactly the kind of life neoliberalism favored, with its dissolution of long-term security and unstable markets. Suffering from poor health, Deleuze leapt to his death from his apartment window in 1995. He was seventy years old.[11]

Objectification

According to Alexander Galloway, control societies bring to life nonliving things.[12] In other words, in modern societies, 'material objects...have a tendency to become aesthetic objects.'[13] Once something becomes aesthetic – that is, once it can be separated from reality and captured in artistic form – it takes on what Galloway calls a 'second nature.' Drained of deeper socio-political meaning, aestheticized objects become 'autonomous, living entities.'[14] They develop into beings with agency that affect human lives. In return, we humans merely gaze back at them and comment on how beautiful, striking, erotic, or pleasant they are. We make up metaphors to describe them. As this happens, objects come to exist for their own sake.

Walter Benjamin warned that politics, too, could become aesthetic. Once this happened, he believed, fascism would inevitably rise. Politics would turn into a harlequinade. Debates over real policies would be replaced with spectacle and buffoonery, leading to a stagnation of progress. True social change for the betterment of the public would halt. Things would mostly remain the same with only the actors changing. Historical events like wars would be primetime entertainment. Some would be seduced by the imagery of war – its remote violence and dualistic narrative – and even come to regard its destruction as splendid.[15] Like war, capitalism too would become aestheticized and take on this second nature. The elegance of markets, the exquisite purity of numbers, the rush of limitless growth – these attributes are enticing to countless free-marketers.

If everything is aestheticized, then everything is an object. Even life itself gets objectified. Once made into a representation, life is considered on equal playing field with machines and matter. Things become interchangeable and reducible to the shared essence of data. Objects are thought of as information to be fed into systems, altered, and ultimately controlled.

But when we think of life as information, for example, we lose a great deal in the process. In *How We Became Posthuman*, media scholar N. Katherine Hayles critiques the notion that information can be divorced from the matter that supposedly houses it. Reduced to data without material dimensions, she writes, humans undergo a radical change. We lose the very experiences that make us 'embodied creatures.'[16]

Once people begin to think of things as information, socio-economic relations shift drastically. Because information can be passed from object to object like a virus, the idea of ownership disappears. No one owns information; instead, information is something that is accessed. 'Information is not a conserved quantity,' Hayles stresses. 'If I give you information, you have it and I do too.'[17] As capitalism spawned control societies across the world, and information wrenched loose from immovable disciplinary structures, many felt liberated from the older ways. Anything we might want to know could be accessed at any given time.

However, Hayles notes that access can also separate the 'haves from the have-nots.'[18] Today, heated debates rage over who is allowed entry into certain spaces and who is kept at the door. The policing of movement from place to place, the raising of border walls, the encryption of data – these and other restrictions lock out scores of people. And monitoring technologies keep watch to make sure only the chosen are let in. But you can't keep information from spilling out, and you can't always manage the chaotic flows of human migration. Somewhere someone will break through. Information will leak. What the state does with those incomers and leakers sets long-term precedents. Things cannot forever be kept under lock and key.

Modern literate cultures function best when things are considered objects. Unlike oral cultures, which rely solely on the human-to-human exchange of words to get things done, cultures that read and write must attend to other objects in the world. They organize around symbols and written words and, over time, reduce the need for face-to-face communication. A natural consequence of literate cultures is that people retreat into themselves. Knowledge acquisition becomes a private affair as people read alone. Individuals are split from communities. Public and private realms are severed.[19]

As highly developed technologies are built, literate cultures structure everything around objects. In Western digital societies, this objectification is necessitated by the invention of object-oriented computer programming, which literate cultures come to realize is more efficient at solving complex problems than older linear programming languages. Object-oriented programming is the computer language of choice for nearly every computerized function today, from corporations conducting transactions to artificial intelligence responding to data input.[20] It is, therefore, the very language of commerce.

We live in an objectified world. Aestheticized under capitalism, politics has become a genre of entertainment, starring celebrities, watched by us. So is war, and some regard the violence with pleasure. Women, too, have long been objectified under what bell hooks calls imperialist white supremacist capitalist patriarchy. Human life is objectified by digital enthusiasts as mere consciousness, which might be uploaded to computer systems and altered; people everywhere await the day to upload their minds to computer simulations. Over time, the relationships between things come to mirror the object-oriented processes of capitalism. There is even a philosophical trend known as object-oriented ontology, which posits that all things – even humans – are merely objects interacting with other objects.[21]

But reducing humans to objects aligns well with our technocratic society, in which algorithms seem to have more agency than actual humans. The fact is, you cannot have a democracy of objects any more than you can divorce consciousness from the human body. Embodied creatures, the human animal is indeed entangled with other objects and perhaps not as separate from the world of things as we once thought. But, as Kenneth Burke writes, we can also 'distinguish [humans] from other animals without necessarily being over-haughty. For what other animals have yellow journalism, corrupt politics, pornography, stock market manipulators, plans for waging thermonuclear, chemical, and bacteriological war?'[22]

Objectification occurs everywhere in control societies, and the result has been disastrous, as nonhuman interactants like Internet bots spread propaganda, affect policy decisions, and influence political elections. Because no one is sure how many bots exist online, we may interact with them and not even know it.[23] As things are turned into objects, and we give those things humanlike agency, then we allow them to exercise control over us, perhaps far more than they should. This can make you question whether you're acting freely or if someone – maybe a bot – is controlling you remotely. Those who objectify the world even give agency to information, which they argue is an entity that wants to be free.[24] For this reason, information is the prized object in a control society. Those who have access to it wield tremendous power.

The Ouroboros

Deleuze took note of the world coming into view in the late twentieth century. He understood that greater and more powerful forms of social control would eventually direct the actions of individuals everywhere. What developed over the following twenty years after he died was a massive turn towards nostalgia as publics yearned for the relative stability of disciplinary societies.

Under control, citizens certainly feel free, but they are managed on the finest of levels. Flexibility under the eye of constant tracking and assessment makes for an exhausting existence. Steady, long-term jobs continue to dwindle as short-term gigs take their place. Individuals must put in the work to brand themselves across media platforms in the desperate contest to land a job. Debt piles on. Though we have nearly everything at our fingertips and can work various side hustles to complement whatever full-time employment comes our way, we are still monitored. The illusion of freedom feels liberating, but the prevailing emotion in much of Western debtor society is deep nostalgia for more stable times.

In the 'Postscript,' Deleuze personifies the disciplinary society as a mole and the control society as a serpent, differentiating one from the other on the basis of their respective monetary systems:

Perhaps it is money that expresses the distinction between the two societies best, since discipline always referred back to minted money that locks gold in as numerical standard, while control relates to floating rates of exchange, modulated according to a rate established by a set of standard currencies. The old monetary mole is the animal of the spaces of enclosure, but the serpent is that of the societies of control. We have passed from one animal to the other, from the mole to the serpent, in the system under which we live, but also in our manner of living and in our relations with others.[25]

Whereas the mole digs through subterranean mazes, trapped by the walls of its burrows, the serpent slithers without hindrance. It slides along the grass, smooth and torsional. And unlike the mole, the serpent can spiral into finer coils and perhaps even swallow its own tail. When it does this, it assumes the form of the mythic ouroboros: the serpent that eats its tail.

If a control society is a serpent, then a nostalgic one is an ouroboros. A symbol of recursion, the ouroboros is a metaphorical reminder that histories repeat and old ideas rarely ever die. To prevent the 'new' from taking shape in a nostalgic society, the 'old' probes the walls of history for weak points and bursts through. New ideas appear to take shape, but very often they are built upon old prejudices that refuse to die. Whether by the corporate impulse to limit competition, the legal efforts to freeze the fair usage of content, or the desperate desire to lock out those ideas that don't conform, cultures under control can get trapped within the circle of the snake. We have transitioned into this new kind of control society, a variation on the theme of the serpent.

By confining all activities to the enclosure of the ouroboros, corporations can keep watch more efficiently. Everything is easily seen within the closed circular border of a nostalgic society. What scans initially as movement is in fact repetition. What looks like complexity is actually categorization: the grouping of individuals into tiny, granular chambers, even chambers of one. Motion is permitted but limited. Things become less and less internally diverse. This is true today for both recommender systems and nation-states. Those permitted on the inside, living in the shadows of erected walls, under watchful eyes, are witness to the fascism that spreads when homogeneity sets in.

The Mirrorworld and the Meatworld

There are some in control societies who dream of exercising their own control over themselves, others, and the world. Their desire is to conjure worlds in which they can govern. This dream, told and re-told throughout history, is presently considered to be realizable thanks to advancements in the field of virtual reality.

André Bazin, film critic and founder of the influential French film publication *Cahiers du cinema*, claimed that cinema is the realization of the human desire to foil death. Tracing film's evolution from the earlier plastic arts, Bazin believed the history of art told the story of humans' quest to capture and freeze life in more perfect representations. He attributed this drive to what he called our 'mummy complex' – an innate desire to ward off death by preserving life.[26] From sculptures to paintings, art history is a series of steps leading to a complete re-creation of reality, which is itself a final solution to death. When photography hit the scene in the nineteenth century, it not only seemed to capture perfectly detailed portraits of reality but, in doing so, also relieved painting from the pressure to re-create reality perfectly. For these reasons, Bazin called the invention of photography 'the most important event in the history of plastic arts.'[27]

Bazin alleged that cinema wasn't just a technical invention. It was also a dream. Its inventors fantasized about representing reality.[28] They were hindered merely by the technologies that would eventually arrive, the techniques and devices that would enable one to preserve a more perfect representation of life on celluloid. Perhaps one could live in that representation and escape the clutches of death. From the beginning, the dream of cinema was the human dream of total immersion.

In our present time, steps are being taken to advance beyond the limitations of cinema. Although in no short supply today, moving images on the screen pale in comparison with the promises of virtual reality: its totalizing capabilities, complete immersion, and limitless possibilities. What cannot be totally mapped in virtual realms is augmented instead: the real and fantastic merging into one.

Virtual and augmented reality enthusiasts dream of creating

97

an interactive universe of one, of which the user is in ultimate control. Like Prospero, the dethroned magical duke in William Shakespeare's *The Tempest*, a single person can oversee every minute detail of a given fantasy world. In the play, Prospero is usurped by his brother and banished to a remote island, where he learns the dark arts. Armed with supernatural powers, he controls every aspect of the island and its inhabitants, including the mooncalf Caliban and the spirit Ariel, whom Prospero frees from a pine tree. Threatened with punishment by the sorcerer, Caliban rebels but Ariel obeys. Like Prospero's magic, virtual reality simulations will permit the usurped, the frustrated, and even the vengeful the power to assert rightful claim and rule over their own private kingdoms. Whoever slights the user in the meatworld will be properly dealt with in his totalitarian mirrorworld.

The consequences of such a *userverse*, in which a single user has absolute domain, have been rendered often in narrative media. Several episodes of Rod Serling's mid-century classic series, *The Twilight Zone* (1959-1964), for example, portray characters that wish for worlds of their own. Upon receiving them, they nearly always go mad.

In 'The Little People,' from *The Twilight Zone*'s third season, two astronauts – Craig and Fletcher – crash land on an alien planet. While Fletcher tends to their spacecraft, Craig, a hotshot with an inferiority complex, happens upon a race of microscopic people and quickly decrees himself their leader. Fletcher mends the ship, but Craig is drunk with power and refuses to depart with him. Fletcher eventually sets off for earth, leaving Craig alone with his tiny kingdom, until two gigantic astronauts appear. Craig screams at them to go away, declaring himself the only god around. One of the giants picks Craig up to get a closer look, crushing him to death in his hand.[29]

'The Little People' is a fable about the self-destructive element of power, but it's also about the lengths a person will go to maintain rule over his userverse. Having played second fiddle to Fletcher, Craig finally gets his chance to be number one. At long last, he feels what it's like to give orders, but his desperate thirst for control ends him. Craig is only a god to the tiny people; to the giants, he is small.

'USS Callister,' the frontrunner episode from the fourth season of the hit anthology series *Black Mirror*, bears several similarities to 'The Little People.' In the episode, a successful but underappreciated video game designer, Robert Daly, creates his own virtual reality simulation that resembles a makeshift representation of *Star Trek: The Original Series*. Stepped on by his co-workers and awkward in his relationships, Daly returns home each night to eat take-out and play his game, in which he rules as a knockoff Captain Kirk. Members of his crew in the game resemble his acquaintances in real life. When co-workers frustrate Daly outside the game, he exacts revenge on their doubles inside it. Under constant fear of torture without the relief of death, the crewmembers in the game live out a meaningless existence as fawning slaves.[30]

'USS Callister' presents the frightening consequences of userverses, especially those created and ruled by power-hungry misogynists. Like 'The Little People,' the episode reveals the grotesque fantasies some will conjure in order to satisfy their desires. A striking difference, of course, between both episodes is that Robert Daly has created his own virtual world in which he exerts near-total control; Craig merely stumbles upon his. Daly's world, however, is completely immersive – a manageable universe that often feels more real than reality.

The quest for total immersion is also one to invent a technology that serves people – to satisfy social needs, deliver constant information, and ease the strain of daily life. Amazon has sought to create such inventions with their smart devices and smart speakers. Like nearly every digital gadget in the consumer electronics marketplace, Amazon's smart speakers raise privacy concerns since their default setting is to always listen in case a command is given to it. The hope for Amazon is to net an enormous profit by providing a product purportedly designed to usher in a utopian singularity where humans are harmlessly aided and catered to by machines.

Bazin would have likely considered virtual reality the logical *telos* in the lineage of the plastic arts. By giving consumers the power to live in and even rule over whatever fabricated world they could desire, tech corporations will ultimately bring about a world in which everyone who can afford a device can have limitless control over simulated surroundings. But if virtual reality will be anything like the Internet, the majority of people will use it to fulfill sexual fantasies, manipulate others, and quell the fear of being alone.

We may not have very long to wait for these things to come to pass. The augmented and virtual reality market is a billion-dollar industry, and interest in it has piqued thanks to affordable head-mounted devices. For those in the industry, virtual reality is the very mark of the future – so much so that the technology has yet to catch up with the dream. Yet virtual reality devices trade more in the market of regressive goods than in any market of 'the future.' Virtual reality promises a world made in our image, our own island where each of us can be Prospero in control of every facet of life. We can return to a more childlike state, where our every need was met and the world was one that revolved around us alone.

But the dream of immersion is not the sole force guiding the development of technology. We may desire manageable virtual userverses, and we may even think the people of the past dreamt of the technology to come (though they were not stupid or lesser for doing so). We might believe, as Bazin did, that we suffer from a mummy complex. But ultimately the desire for more perfect technologies, and virtual reality in particular, is also one to amass enormous wealth. Any tech company that invents an affordable virtual reality device that completely immerses users will make it and its shareholders unfathomably wealthy.

Many in the present century might be pushing towards more complete immersion, but for others, it is as much a pursuit of wealth as a long dream. If we assume we are lesser because we lack immersive control or virtual reality, and that successors are silently pitying us from the future, then we fall victim to the mummy complex. We risk assuming the future has already been determined, and, therefore, we must hurry up and invent it. But no one, not even Big Tech, can predict the future.

And although no virtual reality simulator with this power yet exists, our contemporary technologies make similar promises. By knowing anything we ask can be answered by our smartphone or intelligent personal assistant, we can continue to be taught as children by an all-knowing greater presence. We can give orders like Craig or Robert Daly and live in one virtual fantasy after another, no matter how infantile or deviant they might be. What may seem at first like a fresh pursuit of the future is actually a desperate attempt to slip the chains of the human condition and retreat to some utopian, childlike state – to thwart death and, by extension, escape the bondage of time.

The Final Virtual Solution

André Bazin understood well the allure of realism. He knew humans to be thirsty for the ideal copy, the most perfect representation. What he neglected was our craving for the unreal. It's true, there are some who desperately seek to control time and ward off death. But given the mystical power of creating worlds, many will leave realism behind and summon fantasies instead. When the real world is inadequate, illusion steps in.

There are several fantasy narratives set in the idylls of virtual reality. Some, like *Tron* (1982) and *The Matrix* (1999), portray simulated reality as dangerous but fun, a place where you could be killed unless you learn to perform magnificent physical tricks that defy normal laws of physics. Others, like *Ready Player One* and *Black Mirror*'s season three standout 'San Junipero,' are tales about the wonders of virtual reality. They illustrate the awesome possibilities of virtual reality, where anything imaginable can be done.

Created by Charlie Brooker, *Black Mirror* is a streaming anthology series that addresses the anxieties of twenty-first century digital technology with compelling and often grim stories. Like the best of *The Twilight Zone*, *Black Mirror* tackles several of the most pressing human problems, such as war, isolation, death, and power, but it primarily does so by attending to the numerous ways technology alters what it means to be human. Equal parts satire and warning, Brooker's series illustrates the limits people are driven to by technology and the sickening things some technology allows us to do to each other.

Whereas *The Twilight Zone* often focused on the fantastic, pairing tales about talking dolls with those of alien encounters, all of *Black Mirror*'s stories are set, as one critic wrote, 'in worlds only a few minutes from our own.'[31] Each episode features characters struggling against either alternate versions of contemporary technology, with fictional devices standing in for those that actually exist, or versions of digital technology that *could* exist.

Black Mirror's first season premiered in December 2011 on Channel 4 in the United Kingdom and the second followed in February 2013.[32] The first two seasons featured such memorable episodes as 'The National Anthem,' the story of a British Prime Minister who must have sex with a pig on live television to save a nationally beloved Princess from kidnapping, and 'The Waldo Moment,' in which a cartoon bear runs for public office and, upon winning, ushers in a police state.

In 2014, three years after the first season premiere, Netflix acquired the rights to stream the series and, over a year later, officially bought it.[33] For its third season, *Black Mirror* was rebranded as a Netflix original series. Its acquisition by the streaming giant ensured that a much larger audience would have the opportunity to watch it. Scaling the third season to appeal to more people, and dialing back the intensity of the episodes so that viewers can binge-watch the season, Brooker purposefully wrote fewer 'trap' narratives, in which characters must escape outrageous scenarios and which he felt had dominated the narrative structures of the first two seasons.[34] With less full-throttle stories, the third season of *Black Mirror* lacks the immediate shock of the first two and instead pairs its 'trap' narratives with episodes like 'Hated In The Nation,' a twist on the police crime genre, and 'San Junipero,' the series' first romance.

In a 'conscious decision to change the series,' Brooker wrote 'San Junipero' as a surprise to fans of the show.[35] The episode is a love story between two young women, Kelly and Yorkie, who meet weekly in a virtual seaside town called San Junipero.[36] In the episode, San Junipero functions as a kind of artificial heaven where deceased members can live eternally in youthful bodies. The elderly are allowed to visit the simulation every week for a fixed amount of time, and when they are ready to die, they can decide to 'pass over' into the virtual world. Their consciousness is then uploaded to San Junipero, where they can choose to live forever in whatever simulated time period they want. Kelly and Yorkie leave their real lives and visit San Junipero every week to try out the simulation in their young bodies. Over successive meetings, they fall in love and confess who they really are. In the real world, Yorkie is an elderly quadriplegic, and Kelly is dying of cancer. Yorkie plans to pass over, but Kelly is skeptical. Although she is initially reluctant, Kelly eventually concedes, and she and Yorkie live happily ever after in San Junipero.

'San Junipero' won numerous awards, including two Primetime Emmy Awards.[37] It has proven to be enormously popular for its inclusion of queer characters, its hopeful ending, and its portrayal of previous decades, like the 1980s, 1990s, and early 2000s. Because it foregrounds the generic tendencies of a romance and bucks the criticism that all *Black Mirror* episodes are bleak, 'San Junipero' is a unique entry in the series, yet its depiction of virtual reality as an artificial afterlife is shockingly reductionist. For *Black Mirror*, a series known for promoting a healthy distrust of technology, to offer an episode with such a lack of technological criticism is alarming. That it's proven so popular reveals a great deal about how we in the twenty-first century currently imagine virtual reality.

San Junipero is portrayed as a paradise made possible by invisible digital technology.[38] As they live in the simulation, the characters do not encounter the mighty computing power that would make such a virtual utopia possible. Only at the end of the episode, during the credits, are we shown what kind of server farm could power such a vivid virtual world. We can assume neither Kelly nor Yorkie has seen the colossal server rooms at TCKR, the corporation that invented San Junipero, for they don't exist inside the virtual paradise.

More importantly, the characters in San Junipero live their lives without any digital technologies, such as smartphones, computers, or Wi-Fi, to aid them. The architects of the virtual town have edited out not only any futuristic technologies but also the extant technologies of the twenty-first century that structure much of our own social reality. Because digital devices are absent from San Junipero, the characters communicate without the distraction of ubiquitous digital technology. Lacking the ability to micro-coordinate with one another through global positioning and texting, they are allowed the benefit of getting lost.

The possibility of losing another or even one's self imbues the virtual world with an element of chance. At one point in the episode, Yorkie loses Kelly's whereabouts. Perhaps she loses Kelly forever; what then? A world with constant connection to and accessible information about others is one with little serendipity, but in San Junipero, without specific technologies to alleviate the anxiety of not knowing something, certain things are left to chance and mystery, an idea shocking in our own control society structured by communicative technology. That San Junipero, a virtual world bolstered by technological advancements we can only dream of today, erases actually existing digital devices from its reality seems paradoxical at first blush, but perhaps this erasure speaks to a cultural anxiety. Although we want technology to cater to us, we also want to be free. By relying on such technological advancements to create utopia, the architects of San Junipero manage this tension. The inhabitants of the virtual realm are then able to reap the rewards of digital technology without encountering it.

The people of San Junipero also live without the threat of danger. Presenting San Junipero as a utopia, Charlie Brooker eliminates the possibility of transgression in the simulation. By confining all transgressive acts in San Junipero to a bar at the edge of town called The Quagmire, the episode implies that digital technology can contain dissent and that the basest acts of human depravity will not spill into the wider virtual world.

The brief scene in The Quagmire is the only moment when we are given glimpse into the contained transgressions in San Junipero, but it fails at depicting the kind of abyss some people would actually plumb to test the limits of pain and pleasure in a virtual world. Because Brooker dials back the horror of such a place, he inadvertently reinforces the dangerous idea that virtual reality can quell unspeakable, terrible acts. And because both Kelly and Yorkie think negatively of The Quagmire, we viewers are cued to believe that transgressions in San Junipero are widely frowned upon. Dissent, rebellion, and nonconformity are not welcome in the squeaky-clean artifice of the virtual realm, and even the transgressions that are included in the story, such as snake handling and cage fighting, are relatively tame compared to the worst human cruelties. There are no suicide bombers, rapists, or lynch mobs in San Junipero.

Containing every known transgression in one club means that inhabitants can freely enter into transgressive acts and engage in them consensually with others. Everyone in San Junipero, whether inside The Quagmire or not, can seek pleasure and pain without the fear of coercion. When Yorkie sits atop a building and surveys the downtown of San Junipero, as if she might jump to her death, Kelly remarks that she hopes Yorkie has her 'pain sliders set to zero,' indicating that, if Yorkie chooses to jump, she *could* feel pain in San Junipero. The choice to feel anything at all is ultimately up to the individual. With this choice, inhabitants are given the freedom to live without fear of pain or subjugation. No one harms anyone else in San Junipero, but if you wanted to, you could just visit The Quagmire and consent to it.

Obviously, such transgressive containment would require a management force to ensure order. A virtual simulation like San Junipero would be ruthlessly chaotic without it. Where there would be no dissent, someone would sow it. Yet no police force or military units are shown in San Junipero, meaning that whatever polices San Junipero does so in the shadows. We are shown one instance of this force when Kelly, upset that she is falling in love with Yorkie, punches a bathroom mirror. When she looks up at the smashed mirror, it has been repaired. The overseers, whoever they are, ensure that things are kept orderly and clean in San Junipero, for the sight of broken glass could symbolize lawlessness.

This gesture is rooted in broken windows theory, which posits that law and order can be better maintained if the built environment is tidy.[39] Once building windows are broken, the theory claims, people may tend to break the law more often. The theory also accounts for other unsightly displays, such as public drunkenness, sidewalk trash, or graffiti, that, if witnessed, might nudge society closer to disorder. A tidy society, the theory maintains, deters crime, which in practice has led to a form of police surveillance that unjustly targets minorities and the poor.[40] In San Junipero, broken windows theory has been implemented to deter crime outside The Quagmire, and any act that threatens the known order – vandalism, disturbing the peace, or even just punching a bathroom mirror – is quickly and quietly corrected.

The overseers of San Junipero also bear the burden of policing multiple time periods at once. The setting at the beginning of the episode is San Junipero in 1987, but there are also the San Juniperos of 1980, 1996, and 2002, each of which exists simultaneously. In the virtual utopia, anyone can feasibly travel to any simulated time period they want, thus granting inhabitants further control of their lives.

But the artificial time periods are by no means representative of the actual periods in history they replicate. The simulated 1987, for example, is comprised only of mainstream popular culture artifacts of the time period. The bar at the center of town in each year plays the hits from that time period, so when inhabitants desire a change of music, they can simply hop from one year to the next and revel in the respective pop culture.

We viewers are invited to imagine how wonderful San Junipero must be because the versions of the past it portrays are shorn of war, poverty, sickness, and the like, therefore promising us that virtual reality can serve as an escape from the horrors of history. But there can be no escape from history; there is only the deranged fantasy that one could do so. 'San Junipero' is one such fantasy that functions as a salve for a post-9/11, post-Trump Western society, in which nostalgia for an ahistorical past circulates widely. By erasing the past's troubles, and releasing the characters from the bonds of history, San Junipero is reduced to a tourist attraction wherein inhabitants can play.

A retrofuturistic utopia without pain or boredom is marvelous indeed, but the virtual Eden presented in 'San Junipero' is even more remarkable than that. The episode also imagines a future where humans have solved death and where virtual reality itself functions as a kind of secular heaven. No one really has to die because it is implied that those nearing death can have their consciousness uploaded to San Junipero. Therefore, future people are given a choice: live in the heaven of San Junipero or perish. Who would choose the latter?

Although it reinforces a powerful myth about the role of digital technology in society, 'San Junipero' succeeds as a queer love story. It is a moving tale of true love without boundaries, the kind of love that transcends life and death. It also refrains from killing off its queer characters, a hackneyed narrative trend that has marked LGBTQ+ characters with the curse of death in order to further the stories of other (usually cishet) characters.[41] Yet in spite of this, 'San Junipero' bolsters the belief that one can assume an identity without prejudice in a virtual world. In San Junipero, queer characters can live openly, and Kelly and Yorkie are able to live together forever in a place where vehement intolerance is either managed or confined to The Quagmire. Like *Ready Player One*, in which the virtual reality OASIS allows anyone the ability to form whatever identity they want, 'San Junipero' reinforces the idea that a virtual paradise can be all things to all people and will allow everyone the freedom to self-actualize. This belief may serve as a kind of balm for viewers in the early twenty-first century, in which police brutality, the lack of access to affordable healthcare, theoconservative fundamentalism, and so forth threaten the freedom to live how one wants.

A solution to these and other forms of intolerance is not a virtual fantasy promising to free us of technology, history, and death, but virtual reality narratives like 'San Junipero' serve as such in a society that still believes in the myth of digital utopia, perhaps more so than ever before. Lured by this myth, societies begin to think of technology as a worldview, a secular religion for those who worship at the altar of Big Tech. Its architects, the Silicon Valley technocrats who vow to cure social ills with artificial intelligence, circulate this myth, and popular media serve as mouthpieces through which the myth can more loudly emit.

A virtual world in which the elderly can engage in immersive nostalgia therapy to help with Alzheimer's could prove beneficial, and one in which quadriplegics can walk, even dance as Yorkie does, could be life changing for some. But waiting for these miraculous advancements to cure both disease and disability only places the elderly, disabled, and anyone else in the stream of curative time – *always waiting for the cure to come*.[42] San Junipero is that kind of miracle technology, but it is also a utopia in which advanced technology razes every last human dilemma to the ground, transforming life into one long act of waiting until the day comes to enter into the glory of San Junipero, where anyone can be anything. Even though cyberlibertarians, hacktivists, and the creative team behind 'San Junipero' may preach otherwise, no technology to date has ushered in utopia. Instead of these digital fantasies, we need universal senior care, major transformations of the built environment to accommodate the disabled and the debilitated, and policies that do not discriminate against one's identity, among so many other hopes. A massive nostalgic paradise is merely a fantasy, an older myth updated for our time. Still, the episode reminds us that love is a force nothing can halt, and if 'San Junipero' can start conversations about queerness, bisexuality, ageism, and ableism, then, despite its foregrounding of harmful digital myths, it will have achieved much.

Puppeteering

When a friend joked that he was addicted to *Black Mirror*, I remarked that his addiction to a Netflix streaming series was itself a *Black Mirror* episode. Giant tech corporations that turn major profits by guiding and even predicting human desires sounds a lot like a plot lifted straight from the dystopian series. The behavioral manipulation of Netflix, YouTube, and other streaming sites has now become part of the process of consuming culture, but consumers are becoming increasingly aware of the algorithms that direct and influence their tastes. One way for these corporations to hide their manipulation is to give the illusion of control to viewers.

In 2018, Netflix released a *Black Mirror* movie that allows viewers to choose the course of the plot from a series of pre-determined options. The interactive film, entitled *Bandersnatch*, tells the story of a video game programmer who slowly grows to believe some greater force is controlling him remotely. That force is the viewer, who is prompted throughout the film to make decisions that alter the flow of the movie. Like a choose-your-own-adventure novel, *Bandersnatch* puts the agency of the characters and the fate of the narrative in the hands of the consumer, who, like Netflix itself, controls the actions of others remotely.

But *Bandersnatch* is a movie, with hours of pre-made footage, and the control given to viewers is illusory. Viewers are merely controlling their *own* experience watching the film. There is no real person on the other end being manipulated. Netflix, on the other hand, is an enormous control apparatus that recommends its titles from a fixed set of choices. The viewer, sifting through the near-endless catalogue of titles, is the one controlled. The fantastic scenarios depicted in *Bandersnatch* and much of *Black Mirror* deter users from the reality that, to Netflix, *they* are the product and that intrusive, manipulative technologies are not the stuff of science fiction – they are already here.

In early 2019, technology policy researcher Michael Veale discovered that Netflix records every decision a viewer makes while watching *Bandersnatch*. Made possible via a General Data Protection Regulation request, which gives European citizens the right to access their own data harvested by corporations, the revelation proved that the streaming giant uses *Bandersnatch* data to further personalize its recommendations to viewers.

Although other media companies have acted similarly with user data, Netflix doesn't ask viewers' permission before watching the film, leading some to regard the film as a ploy to mine viewer data more deeply.[43]

Whether or not this is true, *Bandersnatch* has helped to buttress the theory among Westerners that reality is an elaborately conceived and vastly complex computer simulation created by a future software engineer. Famously, entrepreneur and Tesla co-founder Elon Musk has stated that he thinks we live in a simulation, that it's a real possibility given the speed at which computer technology has advanced over the twentieth century. At that rate, it's only natural to assume humans – or perhaps even a more advanced species – could invent technology to create their own realities, one of which is our own world.[44]

These theories almost always reveal a deep desire to hand over control to a larger entity – a god or perhaps a software designer. It is a kind of virtual determinism that popularizes the idea of someone else controlling the strings. Mere puppets, humans no longer have to think or act, for someone more powerful is doing that for them. Technology becomes the means by which humans sew the strings onto their bodies and guide them up into the hands of the controller. It is a hopeless manifestation of the digital sublime, a way to give up the last vestiges of agency in a desperate bid to achieve freedom.

Humans are neither puppets nor unhindered agents. We are free but bump into the constructed world around us. There is no godlike marionettist manning the control bar. Yet norms, structures, capitalism, forms of intolerance, built and natural environments, and other elements constrain our actions.

Playing The Game

It might seem quite natural to assume life is a simulation when much of it runs like a game. Anthropologist David Graeber describes the gamification of life as a result of the massive bureaucracies that cropped up over the last century. In video games, he writes, 'nothing is actually produced, it just kind of springs into being, and we really do spend our lives earning points and dodging people carrying weapons.'[45] A game, then, functions as a 'utopia of rules' in which 'all ambiguity is swept away.' The gamemaster controls us the players, who are simply 'the playthings of destiny and fate.' A simplified model of reality, a game is where we can come up with the rules while bowing to the demands of the ruler. Or the rules are delineated clearly for us, and everyone starts on equal footing in the race to the finale. Considered in these ways, social media are games. So is capitalism.[46]

Once we start to view the world as gameplay, we risk marginalizing the truly remarkable and forgetting the utterly unconscionable elements of reality. Take for example the YouTube channel TierZoo, which cranks out videos explaining the natural world as if it were a video game. Inspired by list systems that rank video game characters based on their abilities, TierZoo titles include 'When Earth Was In Beta,' 'How Humans Broke The Game' and various tier lists of other animals. In the videos, employing the parlance of games, animals can unlock abilities, level up, learn skills, and recharge. Geographic areas are servers; major epochs are patches; and species are builds and factions. 'For those of you looking to optimize your gameplay in real life,' the tagline reads, 'look no further than this channel.'[47]

TierZoo is an extension of the subreddit /r/Outside, a forum in which posters discuss reality as if it were a massive video game, using the very same language that appears in TierZoo. Humans are players and life is a game. Anonymous posters one-up each other to cleverly explain any life event as if it were part of a game. At first, the threads read like extended jokes, but speaking of life in terms of upgrading, learning skills, training, and completing missions aligns well with the corporate speak of neoliberalism that refers to life as a great competition in which only the winners survive.

If one learns the right skills, the thinking goes, then one might win the game of life. This ultimately means achieving bodily perfection, making money, and being happy. Knowing the rules means playing the game well. Players can then manage their own lives in their surroundings. They might even control others and attempt to bend them to their will. When others do not comply, some players may get frustrated or even hostile. They think, *The rest of the world can be controlled, why can't other people?*

This desire rages within male supremacists known as incels, or involuntary celibates, a reactionary group whose name derives from a term first coined to form community among queer individuals.[48] Forged within the most revolting abysses of the manosphere, the Internet echo chamber in which men's rights proponents and various supremacist hate groups thrive, incels regard themselves as lower-tier players in a Darwinian game of misogyny. The higher tier players are pickup artists, who assume hard and fast rules about dating and relationships, endeavor to master the game of seduction, and win the admiration of the women they pursue. To compete with pickup artists, some incels try everything from hacking female psychology to undergoing the knife in a bid to achieve bodily perfection.[49] Others actively hate the pickup artists for successfully mastering the game and winning over women. Incels believe women wield sex to catch the most attractive male mates, leaving large swaths of lesser men on the margins with no one. The more women make gains in society, the more power they hold over men. If women ruled the world, incels believe they would be no match. Only the fittest alpha men would survive in this game of life.

The pickup artist urtext is commonly considered to be Neil Strauss' 2005 book, *The Game*. While other male supremacists adhere to the rules to become the sort of plummy prince described in *The Game*, incels may turn to violence to either hack the game or exact vengeance on the women they believe have wronged them. Like the Robert Daly character in *Black Mirror*'s 'USS Callister,' these misogynists think they can correct those who abuse them by exercising force. They glorify violence and hate, believing freedom can be gained through killing. When they feel the game becomes impossible to master or even play, they simply carry out with hostile actions what starts as hateful rhetoric. Several of these entitled extremists have committed mass murders in the West. Many of them blame women for the violence.

There is a long history of men regarding women as controllable. Because technology is often gendered masculine, and because much of the major technological industries have been dominated by a small class of men, women have been categorized as just another technology to control. They become thought of as one of man's technologies – a device to cook, clean, have sex with, and bear and raise children. While Western man blanketed the world with grids and taxonomies meant to order things, women were quite often relegated to unpaid domestic duties. Anyone not slotted into these intelligible grids couldn't be governed. At best, they were ignored; often they were punished for being nonbinary.

No matter what incels might think, women are neither controllable technologies nor beings of nature whose essential instinct is to bear children. But when you see the world as a game and life as a zero-sum quest, relationship building becomes a winner-take-all competition, and potential partners are ranked based on their desirability. Watching the winners celebrate their victories fuels the losers with hate. When nothing else works, some incels turn their resentment into action. Murdering becomes a rebellion against the game, which they consider rigged by women. Barbarous extremists, these men represent only some of the intolerant supremacist groups collecting online.

For incels, reviving the good old days might prove effective for taking back the power they believe has been stolen by women.

Their ideal world would look something like the whitewashed Americana of 1950s suburban sitcoms or the macho narratives of Reagan-era mainstream media, both remembered as periods in which white men practiced unchecked power. Returning to these times would mean scaling back the gains made by women and minorities and re-suturing binaries. Never mind the fact that history is riddled with masculinity crises, from feelings of depression during the Industrial Revolution to impotence after WWII.[50] If anything, the hetero fantasies of *Leave It To Beaver* and *Rambo* were meant to soothe male crises of their time periods, not reflect the lack of them. Yet men raised on such fantasies come to suppose they represent the way things were. And if you control the past, you're certainly in a better position to control the present.

In these fantasies, society seems manageable, especially to advantaged white men. Things appear to hinge on simple rules that, when followed, only really benefit a privileged few. But escaping from the present and hiding away in a romanticized version of the past is a fool's errand. It cannot be done. Such a desperate attempt to regain what never existed will only destroy the nostalgic individual.

Whether through technology or mass murder, no new order can be fully realized. Nothing will empower the human animal to escape the material world and float freely as data – always in control, always controlling others. Games are manageable, and in being so, they give the illusion of freedom. But it's really only freedom from living in reality.

Conclusion

Carrying the Flame

Technology...the knack of so arranging the world that we
don't have to experience it.
- *Max Frisch*[1]

We want to stay connected, and we want to be free.
- *Jacqueline Olds and Richard Schwartz*[2]

Stories are more than just things we make up. No story ever
functioned merely as a way to pass the time. Even when we
intend them to be simple entertainment – laughing at a movie,
jumping at a campfire tale – stories help create the world. We
cannot live without them, for they aid in the invention of the
future.

But not everything remains a story. If reality were simply a
bunch of myths strung together, then it might be easy to ignore
true suffering. Some believe there to be nothing behind the scrim
of narrative, that material reality is itself a story somebody made
up, that each of us carries our own stories through which we
view this make-believe world. Isolated from each other, they
labor under the illusions that true meaning doesn't exist and real
change can only occur as a result of telling a different story.

This way of thinking ignores the fact that stories are told
by storytellers. Someone benefits from the myths that circulate
through society. Ruling elites ensure that the narratives told and
re-told through time support their moneyed interests. Changing
the narrative helps to rethink the world, unless the storyteller
has different goals in mind. Those in power have the authority
to conjure myths by which the rest of us live. Storytelling is a
privileged art.

In 2016, Google designer Nick Foster made a fictional video and circulated it among his fellow Google employees. Entitled 'The Selfish Ledger,' the video describes a hypothetical service in which Google nudges users towards specific goals based on the data it collects. Inspired by Richard Dawkins' concept of the selfish gene, which posits that organisms are 'survival machine[s] for...gene[s],' the selfish ledger theorizes humans as 'custodians' or 'transient carriers' of our own data.[3] Data, like Dawkins' genes, can guide us towards specific goals, and Google would serve as the mastermind compiling information and leading us towards the goals that please our data.

To carry this out, the video claims, Google will need unprecedented access to our information. If our ledger determines it's missing a key source of information, it would recommend specific products that could provide it. Options include smart scales, watches, headphones, speakers, and the like – all optimized to gather exact user data. If the ledger ascertains that the products in the marketplace are not of a user's taste, the ledger will design one and 3-D print it.[4]

The selfish ledger's ultimate goal is to perfect a given user, a process overseen by Google and its massive harvesting powers. The result is a data-centered model of human being. But why confine humans to a given determinism? If we refuse to define ourselves biologically, why should we determine our behavior quantitatively? Why reduce complex organisms to mere numbers?

The answer is that doing so benefits a company like Google, as well as those manufacturers of products the ledger might recommend to a user. The selfish ledger is not a possible scenario for making the world a better place. It's a horrific form of micromarketing, equal parts predictive analytics and algorithmic determinism. And by claiming to be a fiction, the video escapes implication, posing instead as a hypothetical service for the good of all people. But it is, in fact, an allegory for the current business practices of Big Tech. It distracts viewers from the stark reality that they are already being determined by a kind of selfish ledger prototype.

Efforts to target an individual's preferences are already underway. One-on-one marketing is often disguised by tech companies as an altruistic endeavor meant to provide customers with exactly what they want, sometimes before they know they want it. Catering directly to personal taste fills a consumer with the warm feelings of importance. Armed with recommender algorithms and micromarketing, tech corporations are able to guide users towards specific content, goals, and products – in the same way the selfish ledger would.

The truth is, technology companies cannot accurately predict human beings. We are far too unstable, leaky, and prone to unexpected disruptions. Because we have the capacity to be unpredictable, and as much as we like to think otherwise, we are irreducible to numbers. But tech companies often create models that supposedly explain human behavior and relationships and then test humans against them. Eventually, humans come to be explained in terms of the model, instead of vice versa. Rather than predicting, tech companies actually guide behavior and beliefs to fit the model, securing us in feedback loops that enable so-called prediction.

Shoshana Zuboff has called the business of prediction 'behavioral futures.'[5] Tech companies spend considerable resources investing in behavioral futures, manipulating human behavior to sell products and services. To provide a cover, they simply call the practice 'prediction' and win over consumers with the promise of controlling their own lives. But it is merely an illusion. To the tech companies, humans *are* simply carriers of data, hosts for the selfish ledger. When the user dies, the data survives.

The video's creator, Nick Foster, also runs the Near Future Lab, which specializes in 'design fiction,' or the process of conceptualizing speculative objects that might structure our future lives.[6] Much of what the Near Future Lab does is create fictions that might materialize into real things. Although an allegory, 'The Selfish Ledger' functions as more than pure fiction. Like all narratives, it changes the way we think about ourselves and the world. It influences how we consider humans in relation to technology. It teaches us that Big Tech is in the business of promoting human rights and that digital technology is a force for the improvement of all lives. It teaches us that privacy doesn't exist; that consumption is the pathway towards perfection; and that humans can be understood best through numbers.

The Personal and the Political

A common solution to the psychosocial problems Big Tech has engendered is to simply disconnect from its products. Put down the phone. Deactivate social media accounts. Detox from digital devices. Go off the grid. Buy physical media; no more streaming. Cover your computer camera with a strip of paper. These actions assume that we are to blame for what Big Tech has done, that technologies are neutral tools and we are just fallen creatures incapable of using them.

At a recent university-sponsored event, I listened to a keynote speaker claim that to connect with one another in the twenty-first century we must disconnect from our devices. Look up from your phone and take a moment to watch a sunset, he said. Revel in its beauty and wonder; be in the moment. If we continue mediating our lives through social media, he argued, then we'll miss out on everything great the world has to offer.

He is right. Sunsets are magnificent natural wonders. But his advice ignores the political implications of digital reliance. At no point did the speaker mention anything about the data gathered from social media users, or social media's proliferation of fascism, or the fact that some people can't unplug from their devices for professional reasons. It's as if he believed the world would continue spinning if Big Tech disappeared tomorrow. But late capitalism would not take kindly to the disappearance of Big Tech, which is a direct result of capital's global conquest.

Neglecting the political when discussing the problems of Big Tech perpetuates the private-public split that has haunted Western culture since capitalism's ascent. Just as we must reckon with the private lives of public figures, so too must we understand that Big Tech's products do much more than make us lonely, depressed, and anxious. But currently, it is far more popular to repeat the same claims made by the keynote speaker: to pay much needed attention to the psychological effects of Big Tech while downplaying or simply avoiding larger structural effects. Talking about Big Tech in this way only forms part of the whole picture.

It's true that social media usage can negatively affect well-being. It can make us feel wonderful and terrible at the same time. We may become dependent on it for our happiness, even as it makes us anxious and depressed. But we cannot shed light on the mental health problems of social media without also attending to larger structural issues. We cannot teach young people to climb the corporate ladder while also demanding they put down their phones, for digital technologies often reinforce the very corporate qualities instilled in us all from a young age. To be a self-made baron, one must network tirelessly across social platforms. To be entrepreneurial, one must utilize the promotional benefits of social networks. To sell yourself, you must have an online presence.

Corporate elites will tell you to put down your phone and go on a digital detox because that will likely make you more productive, which means more growth for the private sector. Companies can't have employees scrolling distractedly on their phones; that might impede their job performance. In reality, very few major corporations today could function without employees tethered to their devices. It's quite hard to detox from digital technology when your job depends on it. Without Big Tech, the corporate sector at large would collapse.

We must acknowledge digital addiction while differentiating it from other addictive products like alcohol and cigarettes. Some people might rely heavily on digital tech to get them through the day. Their relationship to social media may become pathological, and their health might deteriorate. They could consider social media a main source of fulfillment or joy, even as it makes them feel terrible. Like other addictive substances, the digital technologies of today have a propensity to trap people in self-validating feedback loops.

At the same time, digital devices and social media are more than mere tools. They may not kill us with the same efficiency as smoking does, but they tend to degrade social relations on an unprecedented level. Just as we cannot ignore the advertising mechanisms powering Big Tobacco, so we must address the societal effects of Big Tech. Without understanding Big Tech's link to intolerance, surveillance, neoliberal logic, and fascism, we will perpetuate the political rot that is already underway in the West. We cannot tear down the attention economy by simply logging off. Doing a digital detox is the equivalent of just saying no; neither are effective.

The Attention Economy 2.0

The Center for Humane Technology, formerly known as Time Well Spent, was founded by ex-Valley technocrat Tristan Harris and a team of nonprofit organizers and former Big Tech computer scientists.[7] Their mission is to educate the public on how present technology harms our mental health, relationships, children, and democracy.

Spearheaded by Harris, the CHT's goal is to design more ethical technologies and end what it calls 'human downgrading,' an umbrella term for the negative effects of the attention economy. Downgrading includes 'shortening attention spans, rewarding outrage over dialogue, addicting children, polarizing the democratic process, and turning life into a competition for likes and shares.'[8] CHT refers to this list as the 'Ledger of Harms.'[9]

Harris is not wrong. The attention economy, created by ruthless capitalists in a neoliberal milieu and structured by artificial intelligence, is destructive and must be stopped. But even though he has addressed the political ramifications of Big Tech, his solutions could inevitably reify the very attention economy he seeks to dismantle.

Harris has called for software companies to establish boundaries. One such boundary is a plugin that would limit our time checking and responding to emails. He has also advocated for a 'Time Well Spent certification,' which would be the tech equivalent of an organic label. Named after Harris' advocacy group, the label would indicate to the user that the technology is healthy for the mind. TWS-certified software would be available at a cost as a premium option. All of these solutions would supposedly help people reach their offline goals by limiting their time spent online.[10]

At a 2019 talk entitled 'Humane: A New Agenda for Tech,' Harris boiled the problems of the attention economy down to three: artificial social systems, overwhelming artificial intelligence, and extractive incentives. To fix these issues, he proposed a 'full stack socioergonomic model' for tech companies to use when designing ethical software. With this model, tech companies could invent apps that encourage users to seek support from others rather than creating memes and falling down conspiracy rabbit holes. Instead of tricking and manipulating individuals as current Internet bots sometimes do, A.I. could serve as our own sidekicks working in our interest to 'protect the limits of human nature,' a role that bears similarity to Nick Foster's selfish ledger. And apps could compete with each other in a race to determine what's best for a given user.[11]

It's unclear how any of this differs from the attention economy Harris fights against. The plan is in actuality a full-scale remodeling of being human based on ethical tech that's supposed to recognize brilliance and facilitate common ground. Of course, this model reveals nothing about the outrage of modern life that merely spills over into the online world. We aren't simply polarized because our emotions are manipulated by social media only. Humane tech will not effectively resist the structures of oppression that lock out the poor and marginalized in real life.

Harris believes it's up to him to provide the key to escaping the mental health problems he was complicit in creating.[12] He and his fellow Valley technocrats initially used technology to 'hack' into the minds of social media users, and now he assumes ethical, certified-healthy technology can stop further mental hijacking. Like the attention economy architects before him, he opts for understanding how tech can improve *us*.

In 2017, Harris' colleague James Williams spoke at The Next Web (TNW) Conference, an event modeled after TED that was created to 'bring insight, meaning – and every now and again, the lulz – to the world of technology.'[13] The purpose of Williams' talk at TNW was to reveal that technologies are at odds with consumers and to chart a path forward for tech companies.

During the talk, Williams claimed that digital technologies distract us from achieving our goals, which, he noted, is 'exactly the opposite of what technology is for.'[14] Only the push for more ethical technology can save society from the evils of our present tech. Although he criticized Big Tech's shameless programming of people, he urged the public to 'continue to reap the benefits of our technologies' by 'support[ing] and affirm[ing] the people who create them because they carry that flame of innovation and creativity that is so core to the human project.'[15] And he claimed that, like him, the Silicon Valley technocrats who addicted the entire world with attention-stealing, dopamine-rewarding technologies are still the trailblazers charged with directing the course of human history. Because technology has fallen into the wrong hands, its terrible power has been unleashed, but an ethical thinker like Williams would be able to harness this awesome power for good.

The idea that Big Tech can invent ethical technologies using the same problematic outsourcing and extractive processes is ridiculous, but this is a popular belief currently circulating in Western discourse. No matter how much they have criticized the attention economy, public figures like Harris and Williams are believers in the redemptive power of the digital sublime. They assume we all want to live our lives mediated through their inventions and that technology is the vehicle by which society can be made better and more democratic. Their goal is to teach us that we can use technology to shape unrealized utopias. But it is clear not everyone will benefit from Big Tech's gifts and that the centralized structures of private tech corporations are often more imperial than emancipatory.

An Old Myth In Our Time

We should be skeptical of receiving the antidote cooked up by those who first offered the poison.[16] These former technocrats are warning us of digital technology's dangers only after working on the ground floor of the largest tech companies today. Claiming they mastered the art of hooking users with just one click, they have become some of the most outspoken voices discussing social media's debilitating mental, social, and political effects.

As journalist Bianca Bosker writes, technocrats like Tristan Harris and James Williams are realizing now the 'unwelcome side effects' of their creations, an 'epiphany [that] has come with…the peace of mind of having several million in the bank.'[17] But they ignore the privilege of their positions. They affirm that the public should not revolt against them or their colleagues in tech, the ones who have been charged with the task of 'carrying the flame of innovation and creativity,' as Williams says.[18] They also ignore other costs of social media – the cultural costs, the mental health risks of manufacturing digital products, and the people who cannot just unplug, leave work behind for a few weeks, and attend detox camps to clear the mind. Their solutions also avoid a larger issue, which is that technology has never succeeded at ushering in a utopia and never will.

The technocrats never talk about capitalism. They do not attack the ideologies at the root of the attention economy, the ones upon which late capitalism is built: market fundamentalism, deregulation, ruthless individualism, and privatization. They speak the language of neoliberalism, privileging the on-the-move individual who has tasks to accomplish and places to be. Goals must be met, and their products will help us meet them. Living well means winning, and according to Harris, at a time when our technology has us by the brain stem, '[w]e have to change what it means to win.'[19] According to them, life is a ceaseless progression towards an ideal state, a transcendent state, wherein everything is organized and in control, nothing causes anxiety, and reality is mediated through great and powerful technology.

The belief that more technology can solve the problems of technology is an old myth in our time. Until they escape this instrumental logic, the technocrats will likely never address the larger political structures of neoliberalism. And as long as these structures remain, the attention economy will as well.

The Necessary Steps

Defying Big Tech means challenging the neoliberal rationality of competition and entrepreneurialism that shapes our lives. It means doing more than just logging off. It means understanding that the problems of Big Tech go all the way down to the algorithms and business models that underpin its digital products.

Facebook, for example, has been considered a 'structural arena through which social life is regulated.'[20] Its code influences how we interact with each other in real life. Yet it does not determine our lives fully. Users can push back and make demands. To some extent, they can also organize on social networks and build solidarity. For all its many problems, social media has given voice to historically marginalized populations and aided in the formation of resistance groups. Movements against police brutality have heavily increased in the digital age. People across the world have mobilized online to fight against the corporate assault on the environment. It cannot be denied that without social media, millions would have likely never heard of the targeted killing of unarmed black people by militarized police forces. And the worldwide Occupy protests could not have occurred without social media.

However, these platforms have also enabled the alt-right – a fringe collective of Internet neo-Nazis, white supremacists, and vehement nationalists – to take center stage. Operating for years at the periphery of legitimate political debate, the alt-right coalesced online thanks to the extremism that drives the attention economy. Coupled with decades of neoconservative marketing paid for by billionaires like Sheldon Adelson and Charles and David Koch, as well as reports from think tanks like The Heritage Foundation and the Cato Institute, the fringe right has sought to delegitimize the scientific consensus on climate change, demonize universal healthcare, fund methodologically unsound scientific studies, and elevate dangerously corrupt candidates in the national consciousness.

We have reached a point at which perhaps the only feasible method to counter Big Tech is to break it up. This will not solve every problem any more than entrusting Big Tech to 'carry the flame' will. But it might contain the destructive force of the Valley before it's too late.

The steps to do this must begin by targeting a key source of Big Tech's power: the belief that humans behave best when they act like markets. The constant promotion of the self, transactional models of friendship, mediated narcissism, kneejerk invective to secure the eyes of users desperate for an information fix – these ideologies were not dreamt up by Silicon Valley technocrats. They flow from a series of policies that allowed major multinational corporations to bloat to outlandish sizes and, with the permission of Western governments, penetrate every area of our social reality. Having deregulated capital markets to a bizarre degree, twentieth century neoliberal ideologues like Milton Friedman, Lewis Powell, Margaret Thatcher, Ronald Reagan, and Bill Clinton, merely opened the door for Big Tech to assume its position of power as the arbiter of social life and help to usher in a control society that even Gilles Deleuze could not have imagined. Breaking up Big Tech requires returning what has been colonized by corporations to the people. Until that day, corporate capitalism will continue stalking the public as ruthlessly as the violent being in *It Follows*.

In this age of Big Tech, the chasm between rich and poor is immense and growing. An experiment in social control, neoliberalism manages this extreme inequality by eliminating from public view those swallowed up in the vortex of poverty. The ones who cannot be controlled are executed publicly by police or thrown into crowding prisons to serve corporations as low-wage laborers. Migrants are criminalized in this nightmare future of climate weirding and austerity. Commerce concentrates in fewer locations on the globe, in major metropolises and industry hubs, leaving in capitalism's wake a desert of ghost towns in which prescription painkillers pool. Adding insult to injury, public figures preach the gospel of self-reliance, a cruel sermon that places blame squarely on those locked out behind fencing or border walls built by capital. Fortified against ecological and economic threats, ruling elites continue to gain access to nearly anything they might want – the information and adoration of others, a distended marketplace of products, and so on. The trick of Big Tech is that their products make the rest of us think we too are privileged to have this same access.

The answer is not for us to hide away in the past, allowing pre-Recession nostalgia to infect democracy. We must counter historical amnesia, which means understanding that the culture industry has much to gain from promoting shades of the past that appeal to intolerant impulses. Painting history as a playground of decade-appropriate fashion, music, and style distracts audiences from more troubling historical realities. At the same time, remembering history as the whitewashed utopia that was presented on screens in previous decades justifies certain actions that might be destructive in the present: criminalization of immigrants, heteronormative policies, environmental deregulation, and the like. The world cannot go back to 1950s suburbia. And it can't reinstate Reagan's 1980s empire of illusion.

But we can take tools from the past to fight for a better future, one not pre-determined by the myth of digital utopia. Arming ourselves with a radical historical memory means remembering that social life wasn't always this way. It also means understanding that major corporate players, in the pursuit of greater wealth, led the march to privatize public services and desecrate democracy. The future we are now living was imagined and fought for by policymakers who waged war on unions, organized labor, and the common good. Thankfully, at each step of the way, social movements banded together to blunt the corporate assault on democracy. Now, with much of our lives influenced by persuasive technologies and our personal data hunted by advertisers, we are faced with new challenges in new arenas.

There is no digital utopia waiting at the end of history for us to bring into being, nor is there anything in the circle of the snake but the heat death of fascism. And no freedom will be gained from desperately digging up nostalgia's corpse; its visage is too hideous. There is instead only the urgent and necessary reckoning with Big Tech.

About the Author

Grafton Tanner is the author of *Babbling Corpse: Vaporwave and the Commodification of Ghosts*. His writing has appeared in *The Nation*, *The Los Angeles Review of Books*, *We Are The Mutants*, and *Components*, amongst other publications. His band Superpuppet released their debut EP, *Museum*, in 2018 on Outer Worlds Records. They are recording a full-length album. Grafton lives in Athens, Georgia.

www.graftontanner.com

Acknowledgements

There are so many brilliant people who helped shape this book. I must first thank Doug Lain, Elizabeth Mossman, Christopher Derick Varn, Dominic James, Ashley Frawley, John Hunt, Beccy Conway, Lisa von Fircks, and the publishing team at Zero Books for seeing the potential in a vague idea I had about digital technology and nostalgia. I also want to thank Roger Stahl, Belinda Stillion Southard, and Thomas Lessl for guidance and feedback and for exposing me to ideas I would have never come across otherwise. I am grateful for Simon Reynolds, who showed me the ropes, Roy Christopher, who provided support, and Alfie Bown, who hoisted me up. Kai Riedl lit the fire under me, and for that I am indebted to him.

Brave voices started the dialogue long before me: Astra Taylor, Naomi Klein, Nicholas Carr, Sherry Turkle, Chris Hedges, John Gray, David Graeber, and Henry Giroux. We would all be lost without them. I also want to thank Vincent Mosco, Genevieve Baumann, Rose D'Amora, Richard Menke, Andrew Thompson, Michael Grasso and K.E. Roberts, Jake Hanrahan, Onyew Kim and everyone at A Cappella Books, Emily McBride, Sean Pritchard and Emma Furman, Terry Tapp, Rachel Watkins, Will Walton, and the team at Avid Bookshop. Some sections of the book were published as previous versions in *The Los Angeles Review of Books* and *The Hong Kong Review of Books*. I owe a debt to these trailblazing outlets for allowing me to write for them.

Conversations with Sammi Rippetoe, Nathan Rothenbaum, Carly Fabian, Daryn Sinclair, Luke Christie, Matt Farmer, and Allie Doherty (who asked about the bots) were a crucial part of the writing process.

I am humbled by my friends: Justin Belk, Andrew Bennett, Michael Buice, Bobby Ferguson, David Hine, Perry Lee, Blake Lewis, Dillon McCabe, Greg Moyer, Alejandro Ortiz, and Philip

Spence. Thank you for your friendship.

I could not have written this without the support of my parents, Debbie and Kevin Tanner, and sisters, Lauren and Ashley. Thank you for instilling in me a restless thirst for learning and a passion for justice. I owe everything to you. My family grew over the past few years: thank you Susan and David Pence, Mary-Drayton and Grant, and Lily.

'Suddenly anything is possible again.' (M.F. 1968-2017)

When days grow dark, Anna, you are there. You are all that is good in this world. I am forever thankful for your love and support.

Grafton Tanner
Athens, Georgia
October 2019

Notes

Epigraphs

1. David Bacon, "Up Against the Open Shop: The Hidden Story of Silicon Valley's High-Tech Workers," Truthout (Mar. 4, 2011), https://truthout.org/articles/up-against-the-open-shop-the-hidden-story-of-silicon-valley-s-high-tech-workers-2/.

2. Gilles Deleuze, "Postscript on the Societies of Control," *October*, vol. 59 (1992): 3-7.

3. Mark Fisher, "The Great Digital Swindle," *Repeater Books Blog*, Repeater Books (March 30, 2016), https://repeaterbooks.com/the-great-digital-swindle-by-mark-fisher/.

Introduction

1. Reese Erlich, "Rare Employee Strike at Silicon Valley Plant," *The Christian Science Monitor* (Nov. 24, 1992), https://www.csmonitor.com/1992/1124/24081.html; Bacon, "Open Shop."

2. Nick Statt, "iPhone Manufacturer Foxconn Plans to Replace Almost Every Human Worker with Robots," *The Verge* (Dec. 30, 2016), https://www.theverge.com/2016/12/30/14128870/foxconn-robots-automation-apple-iphone-china-manufacturing.

3. Charles Duhigg and Keith Bradsher, "How the U.S. Lost Out on iPhone Work," *The New York Times* (Jan. 21, 2012), https://www.nytimes.com/2012/01/22/business/apple-america-and-a-squeezed-middle-class.html.

4. Alexandra Topping, "Apple Factory Worker Kills Himself After Disappearance of Prototype," *The Guardian* (July 22, 2009), https://www.theguardian.com/world/2009/jul/22/apple-worker-suicide-prototype-missing.

5. Casey Newton, "The Trauma Floor," *The Verge* (Feb. 25, 2019), https://www.theverge.com/2019/2/25/18229714/

cognizant-facebook-content-moderator-interviews-trauma-working-conditions-arizona.

6. Rui Fan, et al, "Anger Is More Influential than Joy: Sentiment Correlation in Weibo," *PLOS One*, vol. 9, no. 10 (2014): 1–8.

7. For more on tech disruption under capitalism, see Jonathan Taplin, *Move Fast and Break Things: How Facebook, Google, and Amazon Cornered Culture and Undermined Democracy* (New York, NY: Little, Brown and Company, 2017).

8. Grafton Tanner, *Babbling Corpse: Vaporwave and the Commodification of Ghosts* (Winchester, UK: Zero Books, 2016).

Chapter 1

1. *Holy Motors*, dir. Leos Carax (2012; Paris, France: Pierre Grise Productions, 2013), DVD.

2. Torres, "Skim," track 2 on *Strange Futures* (4AD, 2017).

3. Walter Benjamin, "The Work of Art in the Age of Mechanical Reproduction," in *Illuminations*, ed. Hannah Arendt, trans. Harry Zohn (New York, NY: Schocken Books, 1969): 235.

4. Ibid., 236-237.

5. Erik Ortiz, "Bill O'Reilly Severance: Fox News Host to Get $25 Million," NBC News (Apr. 20, 2017), https://www.nbcnews.com/news/us-news/bill-o-reilly-severance-fox-news-host-expected-get-25-n748916.

6. See Erving Goffman, *The Presentation of Self in Everyday Life* (New York, NY: Doubleday, 1959) and Goffman, *Interaction Ritual: Essays on Face-to-Face Behavior* (NY: Pantheon Books, 1967).

7. John Keane, *Power and Humility: The Future of Monitory Democracy* (Cambridge University Press, 2018): 169.

8. Joshua Meyrowitz, *No Sense of Place: The Impact of Electronic Media on Social Behavior* (Oxford University Press, 1985): 292-3.

9. Ibid., 294.

10. For Truman's horrifying, hayseed cosmology, see *Dear Bess: The Letters from Harry to Bess Truman*, ed. Robert H. Ferrell (New York, NY: W.W. Norton and Company, 1983): 39.

11. Meyrowitz, *No Sense of Place*, 270.

12. Ibid., 270.

13. Fredric Jameson, *The Political Unconscious: Narrative as a Socially Symbolic Act* (Cornell University Press, 1981): 289.

14. Ibid., 299.

15. See Mark Andrejevic, *Infoglut: How Too Much Information is Changing the Way We Think and Know* (New York, NY: Routledge, 2013).

16. George Landow, "Hypertext and Critical Theory," from *Hypertext: The Convergence of Contemporary Critical Theory and Technology* (Johns Hopkins University Press, 1991): 105.

17. Ibid.

18. Rarely was any actual coverage by Fox News and its demiurges devoted to Obama's drone policy, his silent treatment towards the proliferation of black sites in the United States, or his general support of the banks.

19. Jodi Dean, *Publicity's Secret: How Technoculture Capitalizes on Democracy* (Cornell University Press, 2002): 66.

20. Olivia Solon, "'Incel': Reddit Bans Misogynist Men's Group Blaming Women for Their Celibacy," *The Guardian* (Nov. 8, 2017), https://www.theguardian.com/technology/2017/nov/08/reddit-incel-involuntary-celibate-men-ban; Jennifer Earl, "Reddit Bans Popular 'Alt-Right' Subreddit Over Policy Violations," CBS News (Feb. 2, 2017), https://www.cbsnews.com/news/reddit-bans-popular-alt-right-subreddits-over-policy-violations/.

21. Jay Hathaway, "Why Reddit Finally Banned One of its Most Misogynistic Forums," *The Daily Dot* (Nov. 10, 2017), https://www.dailydot.com/unclick/reddit-incels-ban/.

22. An earlier version of this chapter is published as "Digital Detox: Big Tech's Phony Crisis of Conscience," *The Los Angeles*

Review of Books (Aug. 9, 2018), https://lareviewofbooks.org/article/digital-detox-big-techs-phony-crisis-of-conscience/.

23. TNW, "James Williams (Time Well Spent) on Why and How to End the Attention Economy | TNW Conference 2017," Filmed [2017], YouTube video, Posted [Jun 2, 2017], https://www.youtube.com/watch?v=vR9EvD0QfEs&t=832s.

24. Honor Whiteman, "Social media: how does it affect our mental health and well-being?," *Medical News Today*, June 10, 2015, https://www.medicalnewstoday.com/articles/275361.php.

25. Mahita Gajanan, "'I Care Deeply About the Democratic Process.' Mark Zuckerberg Reveals Facebook Election Meddling Plan," *Time* (Sep. 21, 2017), http://time.com/4952391/mark-zuckerberg-facebook-russia-meddling-congress/.

26. Dylan Byers, "Facebook: Russian ads reached 10 million people," *CNNMoney* (Oct. 3, 2017), http://money.cnn.com/2017/10/02/media/facebook-russian-ads-10-million/index.html; Ashley Gold, "Twitter: More than 677,000 U.S. users engaged with Russian troll accounts," *Politico* (Jan. 19, 2018), https://www.politico.com/story/2018/01/19/twitter-users-russian-trolls-437247?cid=apn.

27. Gajanan, "'I Care Deeply;'" Richard Nieva, "Facebook overhauls news feed to focus on friends and family," CNET (Jan. 12, 2018), https://www.cnet.com/news/facebook-overhauls-news-feed-to-focus-on-friends-and-family/#ftag=CAD-09-10aai5b.

28. Stanford Graduate School of Business, "Chamath Palihapitiya, Founder and CEO Social Capital, on Money as an Instrument of Change," Filmed [2017], YouTube video, Posted [Nov. 10, 2017], https://www.youtube.com/watch?v=PMotykw0SIk&t=1481s.

29. Olivia Solon, "Ex-Facebook president Sean Parker: site made to exploit human 'vulnerability,'" *The Guardian* (Nov.

9, 2017), https://www.theguardian.com/technology/2017/nov/09/facebook-sean-parker-vulnerability-brain-psychology.

30. Bianca Bosker, "The Binge Breaker," *The Atlantic* (Nov. 2016), https://www.theatlantic.com/magazine/archive/2016/11/the-binge-breaker/501122/.

31. Ibid.

32. Longinus, "From *On the Sublime*," in *The Rhetorical Tradition: Readings from Classical Times to the Present*, 3rd Edition, ed. Patricia Bizzell & Bruce Herzberg (Boston, MA: Bedford/St. Martin's, 2001): 346-358.

33. Vincent Mosco, *The Digital Sublime: Myth, Power, and Cyberspace* (Massachusetts Institute of Technology, 2004): 23.

34. Ibid.

35. Ibid., 22.

36. Stuart Ewen, *Channels of Desire: Mass Images and the Shaping of American Consciousness* (New York, NY: McGraw-Hill Book Company, 1982): 15-16.

37. Ibid., 15.

38. Mosco, *Digital Sublime*, 24.

39. Ibid.

40. Paul Virilio, *Open Sky*, Trans. Julie Rose (New York, NY: Verso, 1997): 40.

41. James Carey, *Communication as Culture* (Boston, MA: Unwin Hyman, 1989): 123.

42. Ibid., 114.

43. Ibid., 114-5.

44. Ibid., 115.

45. Ibid.

46. Ibid., 116.

47. Ibid., 135.

48. Mosco, *Digital Sublime*, 62.

49. Ibid., 140.

Chapter 2

1. Jennifer Rockne, "Branded: Corporations and our Schools," Reclaim Democracy! (Feb. 2002), http://reclaimdemocracy. org/branded_schools/.

2. Luke Bryan, "Light It Up," track 3 on *What Makes You Country* (Capitol Nashville, 2017).

3. Sherry Turkle, *Alone Together: Why We Expect More from Technology and Less from Each Other* (New York: Basic Books, 2011): 163.

4. Ibid.

5. Chris Hedges, "A Nation of the Walking Dead," Truthdig (Apr. 3, 2017), https://www.truthdig.com/articles/a-nation-of-the-walking-dead/.

6. Ibid.

7. Ibid.

8. Turkle, *Alone Together*, 227.

9. "What Is Captology?," Stanford Persuasive Technology Lab, http://captology.stanford.edu/about/what-is-captology. html.

10. Information on Gorgias taken from Bizzell and Herzberg, *The Rhetorical Tradition*, 42-6.

11. Ibid., 47-55.

12. Information on ancient Chinese rhetoric taken from Xing Lu, *Rhetoric in Ancient China Fifth to Third Century B.C.E.: A Comparison with Classical Greek Rhetoric* (University of South Carolina Press, 1998).

13. Information on Isocrates taken from Bizzell and Herzberg, *The Rhetorical Tradition*, 67-9.

14. Carolyn R. Miller, "The *Polis* as Rhetorical Community," in *Contemporary Rhetorical Theory: A Reader*, 2nd Edition, ed. Mark J. Porrovecchio and Celeste Michelle Condit (New York, NY: The Guilford Press, 2016): 246.

15. Sonja K. Foss and Cindy L. Griffin, "Beyond Persuasion: A Proposal for an Invitational Rhetoric," *Communication*

Monographs, vol. 62 (Mar. 1995): 4-5.

16. Ibid., 6.

17. By 2015, that number had inflated to one and a half billion.

18. Ethan Kross, et al., "Facebook Use Predicts Declines in Subjective Well-Being in Young Adults," *PLOS One* vol. 8, no. 8 (Aug. 2013): 1-6.

19. Jason Dean and Ting-I Tsai, "Suicides Spark Inquiries," *The Wall Street Journal* (May 27, 2010), https://www.wsj.com/articles/SB10001424052748704026204575267603576594936.

20. David Barboza, "iPhone Maker in China Is Under Fire After a Suicide," *The New York Times* (July 26, 2009), http://www.nytimes.com/2009/07/27/technology/companies/27apple.html.

21. Stanford Graduate School of Business, "Chamath Palihapitiya."

22. For more ways we think the mind as machine, see Nicholas Carr, *The Shallows: What the Internet is Doing to Our Brains* (New York, NY: W.W. Norton & Company, Inc., 2011): 23.

23. Katie Lynn Walkup and Peter Cannon, "Health Ecologies in Addiction Treatment: Rhetoric of Health and Medicine and Conceptualizing Care," *Technical Communication Quarterly* 27, no. 1 (January 2018), 108.

24. Ibid., 108-9.

25. Ibid., 109.

26. Lucas J. Youngvorst and Susanne M. Jones, "The Influence of Cognitive Complexity, Empathy, and Mindfulness on Person-Centered Message Evaluations," *Communication Quarterly* 65, no. 5 (November 2017): 552.

27. Ibid.

28. Ibid.

29. Ibid.

30. Ibid., 553.

31. For an excellent analysis of the commodification of mindfulness, see Ronald Purser, *McMindfulness: How*

Mindfulness Became the New Capitalist Spirituality (London, UK: Repeater Books).

32. Stuart Ewen, *Captains of Consciousness: Advertising and the Social Roots of the Consumer Culture* (New York, NY: McGraw-Hill Book Company, 1976): 37.

33. Ibid., 38.

34. Ibid., 38-9.

35. James B. Twitchell, *Adcult USA: The Triumph of Advertising in American Culture* (New York, NY: Columbia University Press, 1996): 12.

36. A version of this section first appeared in Grafton Tanner, "Classroom Management: Simon Sinek, ClassDojo, and the Nostalgia Industry," *The Los Angeles Review of Books* (Jan. 28, 2019), https://lareviewofbooks.org/article/classroom-management-simon-sinek-classdojo-and-the-nostalgia-industry/.

37. Natasha Singer, "Privacy Concerns for ClassDojo and Other Tracking Apps for Schoolchildren," *The New York Times* (Nov. 16, 2014), https://www.nytimes.com/2014/11/17/technology/privacy-concerns-for-classdojo-and-other-tracking-apps-for-schoolchildren.html.

38. A version of this section first appeared in Tanner, "Classroom Management," *The Los Angeles Review of Books*.

39. Jean M. Twenge, "Have Smartphones Destroyed a Generation?," The Atlantic (Sep. 2017), https://www.theatlantic.com/magazine/archive/2017/09/has-the-smartphone-destroyed-a-generation/534198/.

40. "Twenty One Pilots | Chart History: Stressed Out," Hot 100, *Billboard*, https://www.billboard.com/music/Twenty-One-Pilots/chart-history/hot-100/song/898453; "Winners: 59th Annual GRAMMY Awards (2016)," Recording Academy, https://www.grammy.com/grammys/awards/59th-annual-grammy-awards.

41. "Twenty One Pilots: Stressed Out," RIAA.com (last updated

July 31, 2019), https://www.riaa.com/gold-platinum/?tab_active=default-award&ar=Twenty+One+Pilots&ti=Stressed+Out#search_section.

42. Twenty One Pilots, "Stressed Out," track 2 on *Blurryface* (Fueled by Ramen, 2015).

43. Turkle, *Alone Together*, 288.

Chapter 3

1. Sherry Turkle, *Alone Together*, 277.

2. *Minority Report*, dir. Steven Spielberg (2002; Universal City, CA: DreamWorks, LLC. and Twentieth Century Fox Film Corporation, 2002), DVD.

3. Alan Light, "Billboard Woman of the Year Taylor Swift on Writing Her Own Rules, Not Becoming a Cliché and the Hurdle of Going Pop," *Billboard* (Dec. 5, 2014), https://www.billboard.com/articles/events/women-in-music-2014/6363514/billboard-woman-of-the-year-taylor-swift-on-writing-her.

4. Shorey Andrews, "How Jack Antonoff and Taylor Swift Became Pop's Dream Duo," A.Side (Aug. 28, 2017), https://ontheaside.com/music/how-jack-antonoff-and-taylor-swift-became-pops-dream-duo/.

5. Josh Eells, "Taylor Swift Reveals Five Things to Expect on '1989,'" *Rolling Stone* (Sep. 16, 2014), https://www.rollingstone.com/music/news/taylor-swift-reveals-five-things-to-expect-on-1989-20140916; Andrews, "Pop's Dream Duo."

6. Light, "Billboard Woman."

7. Ibid.

8. Julia Neuman, "The Nostalgic Allure of 'Synthwave,'" *Observer* (Jul 30, 2015), https://observer.com/2015/07/the-nostalgic-allure-of-synthwave/.

9. Jon Hunt, "We Will Rock You: Welcome To The Future. This Is Synthwave," *l'etoile* (Apr 9, 2014), https://web.archive.

org/web/20170711192215/http://www.letoilemagazine.
com/2014/04/09/we-will-rock-you-welcome-to-the-future-
this-is-synthwave/.

10. This quote taken from Forêt de Vin's Soundcloud page:
https://soundcloud.com/foretdevin.

11. Stephan Wyatt, "M83: Junk," *PopMatters* (Apr. 19, 2016),
https://www.popmatters.com/m83-junk-2495437507.html;
Harley Brown, "Review: M83's Anthony Gonzalez Finds
His Past, Present, and Future Self's Happy Place on 'Junk,'"
Spin (Mar 28, 2016), https://www.spin.com/2016/03/
review-m83-junk/; T. Cole Rachel, review of *Junk* by M83,
Pitchfork (Apr. 11, 2016), https://pitchfork.com/reviews/
albums/21769-junk/.

12. These include retrowave, outrun, darksynth, synthcore,
and fashwave. No known list of subgenres exists.

13. Penn Bullock and Eli Kerry, "Trumpwave and Fashwave
Are Just the Latest Disturbing Examples of the Far-Right
Appropriating Electronic Music," *Vice* (Jan 30, 2017),
https://www.vice.com/en_us/article/mgwk7b/trumpwave-
fashwave-far-right-appropriation-vaporwave-synthwave.

14. Ibid.

15. Kelly Weill and Kate Briquelet, "Dimitrios Pagourtzis,
Texas Shooting Suspect, Posted Neo-Nazi Imagery Online,"
Daily Beast (May 18, 2018), https://www.thedailybeast.com/
dimitrios-pagourtzis-reportedly-idd-as-santa-fe-texas-
shooting-suspect.

16. Simon Brew, "The 47 movie reboots, remakes and sequels
of 2017," *Den of Geek* (Jan 5, 2017), http://www.denofgeek.
com/uk/movies/2017-movies/46275/the-47-movie-reboots-
remakes-and-sequels-of-2017.

17. "Yearly Box Office – 2017 Worldwide Grosses," *Box Office
Mojo*, https://www.boxofficemojo.com/yearly/chart/?vi
ew2=worldwide&yr=2017&p=.htm; Gwilym Mumford,
"Stephen King's It scares off The Exorcist to become highest-

grossing horror ever," *The Guardian* (Sep 29, 2017), https://www.theguardian.com/film/2017/sep/29/stephen-king-it-the-exorcist-highest-grossing-horror-film-ever.

18. "Grosses," *Box Office Mojo*.

19. "About: The 45th President of the United States Donald J. Trump," *Donald J. Trump*, https://www.donaldjtrump.com/about/.

20. Kelly J. Baker, "Make America White Again?," *The Atlantic* (Mar 12, 2016), https://www.theatlantic.com/politics/archive/2016/03/donald-trump-kkk/473190/; Tom Engelhardt, "What Trump Really Means When He Says He'll Make America Great Again," *The Nation* (Apr 26, 2016), https://www.thenation.com/article/what-trump-really-means-when-he-says-hell-make-america-great-again/; Ronald Brownstein, "Trump's Rhetoric of White Nostalgia," *The Atlantic* (Jun 2, 2016), https://www.theatlantic.com/politics/archive/2016/06/trumps-rhetoric-of-white-nostalgia/485192/; Sarah Pulliam Bailey, "How nostalgia for white Christian America drove so many Americans to vote for Trump," *The Washington Post* (Jan 5, 2017), https://www.washingtonpost.com/local/social-issues/how-nostalgia-for-white-christian-america-drove-so-many-americans-to-vote-for-trump/2017/01/04/4ef6d686-b033-11e6-be1c-8cec35b1ad25_story.html?utm_term=.71e4bb0c8e5e; and countless others.

21. Mary E. Stuckey, "American Elections and the Rhetoric of Political Change: Hyperbole, Anger, and Hope in U.S. Politics," *Rhetoric & Public Affairs*, vol. 20, no. 4 (2017): 685.

22. Ibid.

23. Fredric Jameson, "Postmodernism and Consumer Society," *Amerikastudien* 29, no. 1 (March 1984): 55-73.

24. For a robust history of nostalgia in popular culture, see Simon Reynolds, *Retromania: Pop Culture's Addiction to Its Own Past* (New York: Faber and Faber, Inc., 2011).

25. Washington Post Staff, "Full Text," https://www.
washingtonpost.com/news/post-politics/wp/2015/06/16/
full-text-donald-trump-announces-a-presidential-
bid/?utm_term=.a3a72b366a73.

26. Emma Margolin, "'Make America Great Again' – Who Said
It First?," *NBC News* (Sep 9, 2016), https://www.nbcnews.
com/politics/2016-election/make-america-great-again-who-
said-it-first-n645716.

27. Nick Gass, "Trump: I'm so tired of this politically correct
crap," *Politico* (Sep 23, 2015), https://www.politico.
com/story/2015/09/donald-trump-politically-correct-
crap-213988; Caitlin Dickerson, "Reports of Bias-Based
Attacks Tick Upward After Election," *The New York Times*
(Nov 11, 2016), https://www.nytimes.com/2016/11/12/us/
reports-of-bias-based-attacks-tick-upward-after-election.
html?_r=0.

28. Christopher Mele & Annie Correal, "'Not Our President':
Protests Spread After Donald Trump's Election," *The
New York Times* (Nov 9, 2016), https://www.nytimes.
com/2016/11/10/us/trump-election-protests.html.

29. Matt Broomfield, "Women's March against Donald
Trump is the largest day of protests in US history, say
political scientists," *Independent* (Jan 23, 2017), https://
www.independent.co.uk/news/world/americas/womens-
march-anti-donald-trump-womens-rights-largest-protest-
demonstration-us-history-political-a7541081.html.

30. David French, "The Ferocious Religious Faith of the
Campus Social-Justice Warrior," *National Review* (Nov 23,
2015), https://www.nationalreview.com/2015/11/religious-
zealots-campus-social-justice-warriors/; Matt Vespa,
"Snowflake Alert: State Department Offered Employees
Stress Management Classes After Trump Beat Clinton,"
Townhall (Jan 19, 2017), https://townhall.com/tipsheet/
mattvespa/2017/01/19/snowflake-alert-state-department-

offered-employees-stress-management-classes-after-trump-beat-clinton-n2273437.

31. Svetlana Boym, *The Future of Nostalgia* (New York: Basic Books, 2001): 3.

32. Ibid., xiv.

33. Ibid., 10.

34. Ulrich R. Orth and Steffi Gal, "Persuasive Mechanisms of Nostalgic Brand Packages," *Applied Cognitive Psychology* 28, no. 2 (2014): 161.

35. See Krystine Irene Batcho, "Nostalgia: Retreat or Support in Difficult Times?," *The American Journal of Psychology* 126, no. 3 (2013): 365; Alexa M. Tullett, et al., "Right-frontal cortical asymmetry predicts increased proneness to nostalgia," *Psychophysiology* 52, no. 8 (August 2015): 990-996.

36. Krystine Irene Batcho, "Nostalgia and the Emotional Tone and Content of Song Lyrics," *The American Journal Of Psychology* no. 3 (2007): 362.

37. Altaf Merchant and Gregory M. Rose, "Effects of Advertising-Evoked Vicarious Nostalgia on Brand Heritage," *Journal of Business Research* 66, no. 12 (2013): 2619.

38. Elena Oliete-Aldea, "Fear and Nostalgia in Times of Crisis: The Paradoxes of Globalization in Oliver Stone's Money Never Sleeps (2010)," *Culture Unbound: Journal of Current Cultural Research* 4, no. 3 (2012): 351.

39. Ibid.

40. Ibid., 347, 354.

41. Ibid., 350.

42. Ibid., 348.

43. Ibid., 349.

44. Roland Robertson, *Globalization: Social Theory and Global Culture* (London: Sage, 1992): 155.

45. Ibid., 148.

46. Janice Doane and Devon Hodges, *Nostalgia and Sexual*

Difference: The Resistance to Contemporary Feminism (New York: Methuen, 1987): 3.

47. Ryan Lizardi, *Mediated Nostalgia: Individual Memory and Contemporary Mass Media* (Lanham: Lexington Books, 2015): 19.

48. Boym, *Future of Nostalgia*, xvi.

49. Marita Sturken, "Comfort, Irony, and Trivialization: The Mediation of Torture," *International Journal of Cultural Studies* 14, no. 4 (2011): 425.

50. Ibid.; Robertson challenges these claims when he writes, "[T]here has been a tendency to characterize the contemporary proliferation of 'historical' tourist sites as straightforward nostalgia. While undoubtedly much nostalgia is involved in that...we have to bear in mind that much of contemporary tourism involves considerable reflexiveness" (Robertson, *Globalization*, 162).

51. Robert Jay Lifton, *Superpower Syndrome: America's Apocalyptic Confrontation with the World* (New York, NY: Nation Books, 2003): 159.

52. Ibid., 161.

53. Ibid., 163.

54. Ibid., 164.

55. Ibid., 164, 177.

56. Nick Bilton, "Instagram Quickly Passes 1 Million Users," Bits blog, *The New York Times* (Dec. 21, 2010), https://bits.blogs.nytimes.com/2010/12/21/instagram-quickly-passes-1-million-users/.

57. Anita Balakrishnan and Julia Boorstin, "Instagram Says It Now Has 800 Million Users, Up 100 Million Since April," CNBC (Sep. 25, 2017), https://www.cnbc.com/2017/09/25/how-many-users-does-instagram-have-now-800-million.html.

58. "Instagram Ranked Worst for Young People's Mental Health," The Royal Society for Public Health (RSPH)

(May 19, 2017), https://www.rsph.org.uk/about-us/news/instagram-ranked-worst-for-young-people-s-mental-health.html.

59. "Instagram 'Worst for Young Mental Health,'" BBC News (May 19, 2017), https://www.bbc.com/news/health-39955295.

60. Ian Crouch, "Instagram's Instant Nostalgia," *The New Yorker* (Apr. 10, 2012), https://www.newyorker.com/culture/culture-desk/instagrams-instant-nostalgia.

61. Ibid.

62. Susan Sontag, "In Plato's Cave," in *On Photography* (New York, NY: Picador, 1977): 9.

63. Marshall McLuhan, "The Photograph: The Brothel-without-Walls," in *Understanding Media: The Extensions of Man* (Cambridge, MA: MIT, 1994): 198.

64. Irene Xun, et al., "Slowing Down in the Good Old Days: The Effect of Nostalgia on Consumer Patience," *Journal Of Consumer Research* 43, no. 3 (October 2016): 373.

65. Merchant and Rose, "Effects of Advertising-Evoked Vicarious Nostalgia on Brand Heritage," 2619.

66. Tim Wildschut et al., "Nostalgia: Content, Triggers, Functions," *Journal of Personality and Social Psychology* 9, no. 5 (2006): 989.

67. Batcho "Nostalgia: Retreat or Support in Difficult Times?," 355.

68. See Batcho, "Nostalgia: Retreat or Support in Difficult Times?"; Batcho, "Nostalgia and the Emotional Tone and Content of Song Lyrics"; and Krystine Irene Batcho, et al., "A Retrospective Survey of Childhood Experiences," *Journal Of Happiness Studies* 12, no. 4 (August 2011): 531-545.

69. Batcho, "Nostalgia: Retreat or Support in Difficult Times?," 365.

70. Ibid.

Chapter 4

1. GOP War Room, "Ruth Bader Ginsburg Criticizes Treatment of Kavanaugh During SCOTUS Hearings, "Highly Partisan Show," [Filmed Sep 12, 2018], YouTube video, posted [Sep. 13, 2018], https://www.youtube.com/watch?v=AriOjUfbBrw&t=1s&frags=pl%2Cwn.

2. From "The Culture Industry: Enlightenment as Mass Deception," in *Dialectic of Enlightenment*, trans. John Cumming (Verso, 2016).

3. Adrienne Raphel, "Why Adults Are Buying Coloring Books (for Themselves)," *The New Yorker* (July 12, 2015), https://www.newyorker.com/business/currency/why-adults-are-buying-coloring-books-for-themselves.

4. Priscilla Frank, "A Coloring Book for Adults, Because Everyone Deserves to Unleash Their Inner Creative," *HuffPost* (Mar. 24, 2015), https://www.huffpost.com/entry/johanna-basford_n_6925752; Priscilla Frank, "10 Adult Coloring Books to Help You De-Stress and Self-Express," *HuffPost* (Apr. 21, 2015), https://www.huffpost.com/entry/adult-coloring-books_n_7088048.

5. Turkle, *Alone Together*, 224.

6. Michelle Joni, "Preschool Mastermind," michellejoni.com (Jan. 16, 2015), http://michellejoni.com/preschool-mastermind/.

7. Ibid.

8. Michael Crider, "Pokémon GO Passes 100 Million Play Store Downloads in Just a Month," Android Police (Aug. 8, 2016), https://www.androidpolice.com/2016/08/08/pokmon-go-passes-100-million-play-store-downloads-just-month/.

9. A version of this section first appeared in Grafton Tanner, "Stranger Things and the Nostalgia Industry," *The Hong Kong Review of Books* (Nov. 23, 2016), https://hkrbooks.com/2016/11/23/hkrb-essays-stranger-things-and-the-nostalgia-industry/.

10. Katie Collins, "How Stranger Things Inspired an Entire Album of Stranger Songs," CNET (June 21, 2019), https://www.cnet.com/news/how-stranger-things-inspired-an-entire-album-of-stranger-songs/.

11. Tim Moynihan, "The Stories Behind *Stranger Things'* Retro '80s Props," *Wired* (July 27, 2016), https://www.wired.com/2016/07/stories-behind-stranger-things-retro-80s-props/.

12. Ibid.

13. Amanda C. Coyne, "Woman Found Dead at Gwinnett Place Mall ID'd as Georgia State Student," *The Atlanta Journal-Constitution* (Jan. 10, 2018), https://www.ajc.com/news/local/woman-found-dead-gwinnett-place-mall-georgia-state-student/ZBHV966ZJnEViue9Nqb27L/.

14. Stranger Things, "Coming Soon: The Starcourt Mall! | Hawkins, Indiana," YouTube video, posted [Jul. 16, 2018], https://www.youtube.com/watch?v=aXyju7zFwyE&frags=pl%2Cwn.

15. Ra Moon, "Filming Locations: Where was Everything Sucks Filmed?," Filming Locations by Atlas of Wonders, https://www.atlasofwonders.com/2018/02/everything-sucks-filming-locations.html.

16. Yohana Desta, "From Surge to *Showgirls*: A Guide to the 90s References in *Everything Sucks!*," *Vanity Fair* (Feb. 16, 2018), https://www.vanityfair.com/hollywood/2018/02/everything-sucks-netflix-90s-references-interview.

17. For an analysis of social forgetting, see Guy Beiner, *Forgetful Remembrance: Social Forgetting and Vernacular Historiography of a Rebellion in Ulster* (Oxford University Press, 2018).

18. A.A. Dowd, "David Robert Mitchell on his Striking New Horror Film, *It Follows*," *The A.V. Club* (Mar. 12, 2015), https://film.avclub.com/david-robert-mitchell-on-his-striking-new-horror-film-1798277440.

19. Jessica Conditt, "'Stranger Things 2' Basically Gives

Everyone a Cellphone," Engadget (Oct. 23, 2017), https://www.engadget.com/2017/10/23/stranger-things-2-tech-review/?guccounter=1.

20. Elisabeth A. Sullivan, "Believe in Yesterday," *Marketing News* 43, no. 15 (September 30, 2009): 8.

21. Merchant and Rose, "Effects of Advertising-Evoked Vicarious Nostalgia on Brand Heritage," 2624.

22. Ibid.

23. James Bennett and Stan Lanning, "The Netflix Prize," *Proceedings of KDDCup and Workshop 2007* (Aug. 12, 2007).

24. Ibid.

25. See Arvind Narayanan and Vitaly Shmatikov, "Robust De-anonymization of Large Datasets (How to Break Anonymity of the Netflix Prize Dataset)" (Feb. 5, 2008), https://arxiv.org/abs/cs/0610105.

26. Ryan Singel, "Netflix Spilled Your Brokeback Mountain Secret, Lawsuit Claims," *Wired* (Dec. 17, 2009), https://www.wired.com/2009/12/netflix-privacy-lawsuit/.

27. Alexander R. Galloway, *Protocol: How Control Exists After Decentralization* (Massachusetts Institute of Technology, 2004): 113-4.

28. Ibid., 114.

29. Ibid.

30. Ray Bradbury, *Something Wicked This Way Comes* (New York, NY: Avon Books, 1998): 255.

31. I cite Bradbury with a caveat. Although his work has shaped me over my life, I must also reckon with his public misogyny and racism. For evidence of some of his wretched opinions, see Ken Kelley, "*Playboy* Interview: Ray Bradbury," in *Conversations with Ray Bradbury*, ed. Steven L. Aggelis (University Press of Mississippi, 2004): 150-169.

32. Jennifer Wood, "In Defense of 'Halloween III: Season of the Witch,'" *Complex* (Oct. 30, 2013), https://www.complex.com/pop-culture/2013/10/halloween-3-season-of-the-

witch-defense.

33. Including a footnote following a sentence critiquing footnotes is an irony that's not lost on me: Spencer Perry, "Green & McBride Reveal Why They're Ignoring All the Halloween Sequels," ComingSoon.net (June 8, 2018), https://www.comingsoon.net/movies/features/951167-green-mcbride-reveal-why-theyre-ignoring-all-the-halloween-sequels.

34. Andrew Dyce, "Ready Player One: The COMPLETE Easter Egg Guide," *Screen Rant* (Mar. 29, 2018), https://screenrant.com/ready-player-one-easter-eggs-guide/; Michael Rougeau, "Ready Player One's Ending Explained," *GameSpot* (Apr. 4, 2018), https://www.gamespot.com/articles/ready-player-ones-ending-explained/1100-6457818/.

35. Richard Brody, "Steven Spielberg's Oblivious, Chilling Pop-Culture Nostalgia in 'Ready Player One,'" *The New Yorker* (Apr. 2, 2018), https://www.newyorker.com/culture/richard-brody/steven-spielbergs-oblivious-chilling-pop-culture-nostalgia-in-ready-player-one.

36. Jimmy Champane, "Why Halloween 2018 Had to Kill Halloween 2 – Video Essay," YouTube video, posted [Feb. 6, 2018], https://www.youtube.com/watch?v=NvfNFDOdpqM&t=241s&frags=pl%2Cwn.

37. Robert Hariman and John Louis Lucaites, "Dissent and Emotional Management in a Liberal-Democratic Society: The Kent State Iconic Photograph," *Rhetoric Society Quarterly*, vol. 31, no. 3 (Summer 2001): 21, 5.

38. Ibid., 20.

39. Ibid.

40. Ibid.

41. Ibid.

Chapter 5

1. Lorgia García-Peña, *The Borders of Dominicanidad: Race,*

Nation, and Archives of Contradiction (Duke University Press, 2016): 15.

2. Public Image Ltd, "No Birds," track 10 on *Second Edition* (Warner Records, Inc., 1979).

3. Deleuze, "Postscript," 3.

4. Ibid., 4.

5. Ibid., 5.

6. Gilles Deleuze & Felix Guattari, *A Thousand Plateaus: Capitalism and Schizophrenia*, trans. Brian Massumi (University of Minnesota Press, 1987).

7. Ibid., 24-5.

8. For example, the Israeli Defense Force has weaponized DeleuzoGuattarian theories to terrorize Palestinians. See Eyal Weizman, "Walking Through Walls: Soldiers as Architects in the Israeli-Palestinian Conflict," *Radical Philosophy* 136 (March/April 2006): 8-22.

9. Benjamin Noys, *Malign Velocities: Accelerationism and Capitalism* (Winchester, UK: Zero Books, 2014): 1.

10. Ibid., 7.

11. Adrian Parr, "What Is Becoming of Deleuze?," *The Los Angeles Review of Books* (Nov. 8, 2015), https://lareviewofbooks.org/article/what-is-becoming-of-deleuze/#!.

12. Galloway, *Protocol*, 86.

13. Ibid., 88.

14. Ibid.

15. Benjamin, "Work of Art," 241-2.

16. N. Katherine Hayles, *How We Became Posthuman: Virtual Bodies in Cybernetics, Literature, and Informatics* (The University of Chicago Press, 1999): 29.

17. Ibid., 39.

18. Ibid.

19. Walter J. Ong, *Orality and Literacy: The Technologizing of the World* (London and New York: Routledge, 2002): 66-7.

20. Alexander R. Galloway, "The Poverty of Philosophy:

Realism and Post-Fordism," *Critical Inquiry* 39 (Winter 2013): 349.

21. Ibid. In Galloway's article, he critiques recent trends in philosophy that mirror the structures of post-Fordism. He thinks the recent nonhuman turn of object-oriented ontology (OOO), new materialism, speculative materialism, and the like is aligned with the object-oriented (that is, not linear or procedural) programming of Big Tech and capitalism at large. He sees recent nonhuman turns as "suspect" and "politically retrograde" because, by conceiving reality as nothing more than a flat field of objects, they resemble the "infrastructure of contemporary capitalism" (348). "[T]he contemporary mode of production has a very special relationship with object-oriented computer languages," he writes, "just as one might have said fifty years ago that it has a special relationship with assembly line manufacturing or a hundred years ago with the steam engine" (352). In other words, object-oriented programming is the code on which neoliberal capitalism is built.

22. Galloway then turns his attention to the nonhuman turn and to Quentin Meillassoux in particular, who was one of the first philosophers to tackle correlationism. Correlationism is the belief that thought can't get outside itself and that your thoughts are your own (354). He writes that correlationism was all the rage during the height of postmodernism, when scholars routinely attacked the idea that a subject can access reality objectively. Contrarily, anti-correlationists, like Meillassoux, believe there to be a hard reality that humans can access but may not fully understand. Furthermore, anti-correlationists understand thought to be its own matter. Indeed, thought is not human but another object in the flattened ontology of things (ibid.).

23. Galloway thinks reducing the human to another object among many aligns well with our post-Fordist society, in

which algorithms seem to have more agency than actual humans. And, in fact, by disentangling the ontological from the political, by assuming everything is an object and we are too, then we give up projects of emancipation (358). He doesn't consider the nonhuman turn a productive philosophical project to counter oppressive systems in the name of socioeconomic justice. Nevertheless, he notes, our object-oriented democracy is already here, and it is not working. In such a "democracy," the market rules. Competition and rivalry determine worth. And computer software oils the gears of sweatshops and outsourced factories (362-3). "What kind of world is it in which humans are on equal footing with garbage?" he asks. "What kind of world is it in which the landscape is a chaotic nothing-world, unfounded at its core and motivated by no necessary logic...or by the logic of the market?" (364-5).

24. Kenneth Burke, "From *Language as Symbolic Action*," in *The Rhetorical Tradition*, ed. Bizzell & Herzberg, 1344.

25. Emilio Ferrara, et al, "The Rise of Social Bots," *Communications of the ACM*, vol. 59, no. 7 (July 2016): 96-104.

26. Joshua Gans, "'Information Wants to be Free': The History of that Quote," Digitopoly (Oct. 25, 2015), https://digitopoly. org/2015/10/25/information-wants-to-be-free-the-history-of-that-quote/.

27. Deleuze, "Postscript," 5.

28. André Bazin "The Ontology of the Photographic Image," in *What Is Cinema? Vol. 1*, trans. Hugh Gray (University of California Press, 2005): 9.

29. Ibid., 16.

30. André Bazin, "The Myth of Total Cinema," in *What Is Cinema? Vol. 1*, trans. Hugh Gray (University of California Press, 2005): 18.

31. Rod Serling, "The Little People," *The Twilight Zone*, season 3, episode 29, CBS (Mar. 30, 1962).

32. Charlie Brooker, "USS Callister," *Black Mirror*, season 4, episode 1, Netflix (Dec. 29, 2017).

33. Alan Sepinwall, "Review: Brilliant British sci-fi drama 'Black Mirror' comes to Netflix," Uproxx (Dec 10, 2014), https://uproxx.com/sepinwall/review-brilliant-british-sci-fi-drama-black-mirror-comes-to-netflix/.

34. Charlie Brooker, "Charlie Brooker: the dark side of our gadget addiction," *The Guardian* (Dec 1, 2011), https://www.theguardian.com/technology/2011/dec/01/charlie-brooker-dark-side-gadget-addiction-black-mirror; "Black Mirror: Season 2," *Rotten Tomatoes* (2013), https://www.rottentomatoes.com/tv/black_mirror/s02.

35. Todd Spangler, "Netflix Run Brings U.K.'s 'Black Mirror' Into Light for U.S. Auds," *Variety* (Dec 18, 2014), https://variety.com/2014/tv/news/netflix-run-brings-u-k-s-black-mirror-into-light-for-u-s-auds-1201380236/; Debra Birnbaum, "Netflix Picks Up 'Black Mirror' for 12 New Episodes," *Variety* (Sep 25, 2015), https://variety.com/2015/digital/news/netflix-black-mirror-new-episodes-1201602037/.

36. Joseph Bien-Kahn, "What Happens When *Black Mirror* Moves Beyond Traps? It Gets Even Better," *Wired* (Oct 21, 2016), https://www.wired.com/2016/10/black-mirror-traps/.

37. Louisa Mellor, "Black Mirror series 3 interview: Charlie Brooker and Annabel Jones," *Den of Geek!* (Oct 19, 2016), https://www.denofgeek.com/uk/tv/black-mirror/44587/black-mirror-series-3-interview-charlie-brooker-and-annabel-jones.

38. Alex Watson, "'I was really worried about San Junipero' – Charlie Brooker on pushing the Black Mirror envelope," iNews (Aug 25, 2017), https://inews.co.uk/culture/television/black-mirror-charlie-brooker-san-junipero/.

39. "69th Emmy Award Winners," The Academy of Television Arts & Sciences, https://www.emmys.com/news/awards-news/69th-emmy-award-winners.

40. From here, all description and quotes attributed to the episode are from the following citation: Charlie Brooker, "San Junipero," *Black Mirror*, season 3, episode 4, Netflix (Oct 21, 2016).

41. James Q. Wilson & George L. Kelling, "Broken Windows: The police and neighborhood safety," *The Atlantic* (March 1982), https://www.theatlantic.com/magazine/archive/1982/03/broken-windows/304465/.

42. Sarah Childress, "The Problem with 'Broken Windows' Policing," *Frontline* (June 28, 2016), https://www.pbs.org/wgbh/frontline/article/the-problem-with-broken-windows-policing/.

43. "GLAAD – Where We Are On TV Report – 2016," GLAAD (2016-2017), https://www.glaad.org/whereweareontv16.

44. Alison Kafer, *Feminist, Queer, Crip* (Bloomington, IN: Indiana University Press, 2013): 27, 44.

45. Matthew Gault, "Netflix Has Saved Every Choice You've Ever Made in 'Black Mirror: Bandersnatch,'" *Vice* (Feb. 12, 2019), https://www.vice.com/en_us/article/j57gkk/netflix-has-saved-every-choice-youve-ever-made-in-black-mirror-bandersnatch; Jon Porter, "Netflix Records All of Your Bandersnatch Choices, GDPR Request Reveals," *The Verge* (Feb. 13, 2019), https://www.theverge.com/2019/2/13/18223071/netflix-bandersnatch-gdpr-request-choice-data.

46. Anthony Cuthbertson, "Are We Living in a Computer Simulation? Scientists Prove Elon Musk Wrong," *Newsweek* (Oct. 4, 2017), https://www.newsweek.com/are-we-living-computer-simulation-scientists-prove-elon-musk-wrong-677251.

47. David Graeber, *The Utopia of Rules: On Technology, Stupidity, and the Secret Joys of Bureaucracy* (Brooklyn, NY: Melville House, 2016): 34.

48. Ibid., 191-3.

49. TierZoo, "Welcome to TierZoo," YouTube video, posted [Sep. 5, 2017], https://youtube.com/Tierzoo.

50. Peter Baker, "The Woman Who Accidentally Started the Incel Movement," *Elle* (Mar. 1, 2016), https://www.elle.com/culture/news/a34512/woman-who-started-incel-movement/.

51. Alice Hines, "How Many Bones Would You Break to Get Laid?," *The Cut* (May 28, 2019), https://www.thecut.com/2019/05/incel-plastic-surgery.html.

52. Ann Foster, "White Whine: The Past that Incels Long for Never Existed," *Bitch* (June 11, 2018), https://www.bitchmedia.org/article/incels-white-supremacists-past-never-existed.

Conclusion

1. From *Bartlett's Familiar Quotations*, Seventeenth Edition, ed. Justin Kaplan (Boston, New York, and London: Little, Brown and Company, 2002): 786.

2. Jacqueline Olds and Richard S. Schwartz, *The Lonely American: Drifting Apart in the Twenty-First Century* (Boston, MA: Beacon Press, 2000): 1.

3. Vlad Savov, "Google's Selfish Ledger is an Unsettling Vision of Silicon Valley Social Engineering," *The Verge* (May 17, 2018), https://www.theverge.com/2018/5/17/17344250/google-x-selfish-ledger-video-data-privacy.

4. Ibid.

5. See Shoshanna Zuboff, *The Age of Surveillance Capitalism: The Fight for a Human Future at the New Frontier of Power* (New York, NY: PublicAffairs, 2019).

6. "What We Do," Near Future Laboratory, http://nearfuturelaboratory.com/#whatwedo.

7. A version of this section first appeared in Tanner, "Digital Detox," *The Los Angeles Review of Books*.

8. For more information, visit the Center for Humane Tech's

website: https://humanetech.com.

9.   "Ledger of Harms," Center for Humane Tech (Dec. 14, 2018), https://ledger.humanetech.com.

10. Bosker, "Binge Breaker," *The Atlantic*.

11. Center for Humane Technology, "Humane: A New Agenda for Tech," Vimeo video, posted [Apr. 25, 2019], https://vimeo.com/332532972.

12. Bosker, "Binge Breaker," *Atlantic*.

13. "About TNW," *The Next Web*, The Next Web B.V., https://thenextweb.com/about/.

14. TNW, "James Williams," YouTube.

15. Ibid.

16. A version of this section first appeared in Tanner, "Digital Detox," *The Los Angeles Review of Books*.

17. Bosker, "Binge Breaker," *Atlantic*.

18. TNW, "James Williams," YouTube.

19. Bosker, "Binge Breaker," *Atlantic*.

20. Rena Bivins, "The Gender Binary will not be Deprogrammed: Ten Years of Coding Gender on Facebook," *New Media & Society*, vol. 19(6) (2017): 881.

CULTURE, SOCIETY & POLITICS

The modern world is at an impasse. Disasters scroll across our smartphone screens and we're invited to like, follow or upvote, but critical thinking is harder and harder to find. Rather than connecting us in common struggle and debate, the internet has sped up and deepened a long-standing process of alienation and atomization. Zer0 Books wants to work against this trend. With critical theory as our jumping off point, we aim to publish books that make our readers uncomfortable. We want to move beyond received opinions.

Zer0 Books is on the left and wants to reinvent the left. We are sick of the injustice, the suffering, and the stupidity that defines both our political and cultural world, and we aim to find a new foundation for a new struggle.

If this book has helped you to clarify an idea, solve a problem or extend your knowledge, you may want to check out our online content as well. Look for Zer0 Books: Advancing Conversations in the iTunes directory and for our Zer0 Books YouTube channel.

Popular videos include:

Žižek and the Double Blackmain

The Intellectual Dark Web is a Bad Sign

Can there be an Anti-SJW Left?

Answering Jordan Peterson on Marxism

Follow us on Facebook
at https://www.facebook.com/ZeroBooks and Twitter at https://
twitter.com/Zer0Books

Bestsellers from Zer0 Books include:

Give Them An Argument
Logic for the Left
Ben Burgis
Many serious leftists have learned to distrust talk of logic. This is
a serious mistake.
Paperback: 978-1-78904-210-8 ebook: 978-1-78904-211-5

Poor but Sexy
Culture Clashes in Europe East and West
Agata Pyzik
How the East stayed East and the West stayed West.
Paperback: 978-1-78099-394-2 ebook: 978-1-78099-395-9

An Anthropology of Nothing in Particular
Martin Demant Frederiksen
A journey into the social lives of meaninglessness.
Paperback: 978-1-78535-699-5 ebook: 978-1-78535-700-8

In the Dust of This Planet
Horror of Philosophy vol. 1
Eugene Thacker
In the first of a series of three books on the Horror of Philosophy,
In the Dust of This Planet offers the genre of horror as a way of
thinking about the unthinkable.
Paperback: 978-1-84694-676-9 ebook: 978-1-78099-010-1

The End of Oulipo?
An Attempt to Exhaust a Movement
Lauren Elkin, Veronica Esposito
Paperback: 978-1-78099-655-4 ebook: 978-1-78099-656-1

Capitalist Realism
Is There no Alternative?
Mark Fisher
An analysis of the ways in which capitalism has presented itself
as the only realistic political-economic system.
Paperback: 978-1-84694-317-1 ebook: 978-1-78099-734-6

Rebel Rebel
Chris O'Leary
David Bowie: every single song. Everything you want to know,
everything you didn't know.
Paperback: 978-1-78099-244-0 ebook: 978-1-78099-713-1

Kill All Normies
Angela Nagle
Online culture wars from 4chan and Tumblr to Trump.
Paperback: 978-1- 78535-543-1 ebook: 978-1-78535-544-8

Romeo and Juliet in Palestine
Teaching Under Occupation
Tom Sperlinger
Life in the West Bank, the nature of pedagogy and the role of a
university under occupation.
Paperback: 978-1-78279-637-4 ebook: 978-1-78279-636-7

Ghosts of My Life
Writings on Depression, Hauntology and Lost Futures
Mark Fisher
Paperback: 978-1-78099-226-6 ebook: 978-1-78279-624-4

Sweetening the Pill
or How We Got Hooked on Hormonal Birth Control
Holly Grigg-Spall
Has contraception liberated or oppressed women?
Sweetening the Pill breaks the silence on the dark side of hormonal
contraception.
Paperback: 978-1-78099-607-3 ebook: 978-1-78099-608-0

Why Are We The Good Guys?
Reclaiming your Mind from the Delusions of Propaganda
David Cromwell
A provocative challenge to the standard ideology that Western
power is a benevolent force in the world.
Paperback: 978-1-78099-365-2 ebook: 978-1-78099-366-9

The Writing on the Wall
On the Decomposition of Capitalism and its Critics
Anselm Jappe, Alastair Hemmens
A new approach to the meaning of social emancipation.
Paperback: 978-1-78535-581-3 ebook: 978-1-78535-582-0

Enjoying It
Candy Crush and Capitalism
Alfie Bown
A study of enjoyment and of the enjoyment of studying. Bown
asks what enjoyment says about us and what we say about
enjoyment, and why.
Paperback: 978-1-78535-155-6 ebook: 978-1-78535-156-3

Color, Facture, Art and Design
Iona Singh
This materialist definition of fine-art develops guidelines for
architecture, design, cultural-studies and ultimately social
change.
Paperback: 978-1-78099-629-5 ebook: 978-1-78099-630-1

Neglected or Misunderstood
The Radical Feminism of Shulamith Firestone
Victoria Margree
An interrogation of issues surrounding gender, biology,
sexuality, work and technology, and the ways in which our
imaginations continue to be in thrall to ideologies of maternity
and the nuclear family.
Paperback: 978-1-78535-539-4 ebook: 978-1-78535-540-0

How to Dismantle the NHS in 10 Easy Steps (Second Edition)
Youssef El-Gingihy
The story of how your NHS was sold off and why you will have
to buy private health insurance soon. A new expanded second
edition with chapters on junior doctors' strikes and government
blueprints for US-style healthcare.
Paperback: 978-1-78904-178-1 ebook: 978-1-78904-179-8

Digesting Recipes
The Art of Culinary Notation
Susannah Worth
A recipe is an instruction, the imperative tone of the expert, but this constraint can offer its own kind of potential. A recipe need not be a domestic trap but might instead offer escape – something to fantasise about or aspire to.
Paperback: 978-1-78279-860-6 ebook: 978-1-78279-859-0